ALGEBRA
IN
WORDS

*A Guide of Hints, Strategies
and Simple Explanations*

Gregory P. Bullock, Ph.D.

2014

Bullock, Gregory P.
Algebra in words: a guide of hints, strategies and simple explanations

ISBN-13: 978-1499701555

ISBN-10: 1499701551

ASIN: B00JDA8FBE

MATHEMATICS/Algebra/General

STUDY AIDS/Study Guides

First Edition

The United States of America

TABLE OF CONTENTS

INTRODUCTION

What Is This Book?

This book is a guide of common math and algebra topics that are explained in a non-traditional way. It is *not* a textbook, nor is it a conventional study guide. This is a book where basic mathematical and algebraic topics are explained in laymen's terms, sometimes even purposefully redundant terms, to make your understanding easier and your learning curve faster. It's more of a guide of supplemental information and perspectives on the math you must learn.

I tutored Calculus for the Math Department in undergraduate school as a part-time job, and then began teaching math at the collegiate level (Basic Math and Arithmetic through College Algebra/Pre-Calc I) in 2009 to a wide range of students of various ages and math education backgrounds. During that time, I began noticing trends among my students and classes. One major trend I noticed was the divide among people who seemed to "get it," and those who didn't "get it" as easily, as quickly, or in the same way as those who did. Although it's pointless to classify students into groups, my job as an instructor is to help bridge the gap and find mechanisms to help *everyone* "get it."

As my teaching style evolved, I noticed that a lot of math (either in the books or traditional lectures) was taught in a sort of "math language," meaning mostly in numbers, variables and lines of equations and simplification steps... which is all well and good, because that's what math is. But I found that much of it was left to interpretation, which some would *get* and some wouldn't. So I started translating the *math into worded steps* and notes and found that students responded positively to it. This was the bridge over the gap I was looking for. Since then, I began giving explicitly worded notes including, but not limited to, step by step instructions. Through observing common learning patterns among students, I also was able to predict common questions or areas of confusion, so I would give notes to preemptively answer questions such as

"What do I look for?" or,
"When should I use this?" or,
"What will it look like?"

and prepare students for frequent mistake areas by also showing what *not to do*, along with what to do.

These experiences inspired me to record my notes and make them available to any student who wishes they had another resource to make learning math and algebra easier. As I stated, this book is not a textbook, and by that I mean I don't give extensive examples and practice questions the way textbooks do. Math textbooks are generally very good at giving them and contain a wealth of information. But traditional textbooks also teach in a very rigid and often bottom-up way. I've found that many textbooks teach certain topics to such a sub-categorical level of detail that students lose sight of how it connects to the bigger picture. So what I offer are other perspectives to the math from the textbooks, and I sometimes unveil them in a more top-down way.

I believe that:
- giving multiple perspectives,
- giving detailed, step-by-step instructions,
- showing examples with commentary,
- connecting key topics,
- translating math terminology,
- answering frequently asked questions,
- highlighting common mistakes, and
- drawing attention to some of the more minute (yet important) details which

sometimes slip under students' radar, will expedite your ability to realize and absorb material. This will help you get better grades and save you valuable time, frustration, and even money (if it means you don't have to repeat a course). Sometimes students learn faster by seeing things *in words*.

In 2010, I published my first book *GRADES, MONEY, HEALTH: The Book Every College Student Should Read*, which is a collection of advice and stories on a variety of college-related topics geared towards helping students excel and get the most out of their college experience. In it, I dedicated 10 short chapters to how to get the best grades. In one of those chapters, I explain how important and helpful it is to listen to the *words* of your professors, and how taking notes on what they *say* is sometimes better than just trying to quickly record what they put on the board. That's the basis of this book: Putting the *math into words*.

When I was writing *GRADES, MONEY, HEALTH*, I originally intended to include a chapter on some basic math and algebra hints, but my amount of material kept growing, and I wanted to keep the book as short as possible, so I kept the math segments out. Then my math-help-notes kept growing and ultimately developed into its own book... *this* book, dedicated entirely to math and helping any and every math student.

This is *Algebra in Words*.

Why Do You Need Algebra?

Many students, especially those whose majors are not math or science, ask the questions, "Why do I need algebra?" or, "When am I ever going to use this?" These are completely valid questions. On one hand, that question can be asked about any general education class which does not seem to relate to one's major. The general stock answer is: Because it makes you more well-rounded. And that's true, but for many, that answer is still vague and unacceptable.

With regards to math specifically, many instructors put forth great effort and creativity to exhibit reasons and scenarios as to how math is used in everyday life. I applaud any and all instructors who can convincingly convey these answers. Unfortunately, most do not have the answers, and even more unfortunately, they tend to be:

- Will help you with money and personal finances,
- Will help you with measurements around the house,
- Will help you be a more efficient shopper,
- You can help a child with their homework,
- Will help you understand time signatures in music,
- Will tell you how long it will take you and a friend to split a job,
- Etc.

These answers are cliché and unsatisfactory. These tasks require little to no algebra (although they do require an understanding of division, fractions, and decimals). But the truth is: there are resources (like smart-phone apps, or just a classic calculator) that do most, if not all of those functions for us. Another truth is: Those who really need and use algebra for their careers are a select few. However, that doesn't mean you have no use for it.

Here's the better answer to your question: Learning algebra makes you better at problem solving. It also makes you aware that there is order to the world we live in. Algebra is basically a series of rules; you might even consider it its own language. Everything in our lives relate to understanding rules to help us solve problems and navigate through life. That includes:

- Laws, the legal system, and contracts;
- rules, regulations, and strategies of sports and games;
- negotiations;
- learning a new language;
- critical reading;
- computers, smart devices and electronics;
- managing and working with people;

- giving, taking and following directions;
- reading graphs, charts and data tables;
- accomplishing tasks and goals of all sizes;
- and yes, even fixing things around the house.

Is it all solving for "x"? Obviously not, or at least not literally, but we can all benefit from learning how to navigate through rules (and) to solve problems. Algebra will help you be more logical and see life more logically.

There's one more thing. Upon learning and passing algebra, you will feel inspired and empowered. Algebra can be a complicated subject. If you conquer it, you will then have the confidence to take on other seemingly complicated obstacles.

Regardless, algebra is a subject you *have* to take. No one says you have to like it, but you might as well accept it and do your best to succeed in it. This book will help you through it.

Throughout the book, I often refer you to other related topics, so I've included a detailed table of contents (and hyperlinks in the digital format of the book, which can simply be touched or clicked) to allow you to quickly jump to such referred sections. Also throughout this book, I use the variable "x" as the universal variable, even though many problems you will encounter (in class or in life) may feature an unknown variable different than "x".

REVIEW OF THE BASICS

First, we must look at a few basics that will be used and referred to throughout this book. From year to year and class to class, you may have grasped the majority of the material you've learned, and built a good foundation. But, say, over a summer or holiday break, or just from not using it enough, you may have forgotten a few of the more obscure principles. These are here to quickly bring you back up to speed.

The Real Order of Operations: GEMA

You may remember that when adding terms, you will get the same sum regardless of the order in which you add. You may also remember that the same applies to multiplication: You will get the same product regardless of the order of the factors you multiply. These are "commutative properties." However, this does not apply to subtraction or division.

For subtraction and division, *order matters*. Order matters, including when subtraction and/or division is mixed in with terms being added and/or multiplied. Since *order matters*, there are a set of rules in place to help us calculate numbers and terms in the proper order, and to put consistency into the way we do math. These are the *order of operations*. Often, books or instructors do not teach this completely correctly. Here, it will be explained completely, with nothing left to be misinterpreted.

1. Simplify inside Groups first, if possible, from inner to outer. A group is a set of (parentheses), [brackets], {braces}, overall numerators, overall denominators, radicands, and absolute value groups.
2. Exponents *or* roots, whichever comes first, from left to right.
3. Multiplication *or* division, whichever comes first, from left to right.
4. Addition *or* subtraction, whichever comes first, from left to right.

The Truth about PEMDAS

Many students are taught the acronym and mnemonic device PEMDAS, which stands for "Parentheses, Exponents, Multiplication, Division, Addition, Subtraction." I must warn to you be careful of PEMDAS; it is misleading and incomplete. If you learn to rely on it, it can fail you. Letter by letter, here's why:

P: The first order is *groups*, of all sorts, as described above. If you think of parentheses only, and you get to other groups, you might think it applies to parentheses only. A more appropriate first letter and word should be G for Groups. Groups include, but are not limited to parentheses.

E: This makes you think of exponents only instead of roots as well. This isn't a big deal, since radicals can be converted into exponent form (when they are, they are called rational exponents), but they are often in root or radical form, so you must be prepared for that. When both exponents and roots appear in an equation, do whichever comes first, from left to right. Also, remember that radicands should be simplified first, if necessary, as they are technically a group, as mentioned in the first step.

MD: The reason this is misleading is because some people interpret this to be chronologically literal. In other words, some see M before D and think multiplication must happen before division, but in fact, it means that any Multiplication *or* Division come before any Addition or Subtraction. But multiplication or division should be treated with the *same priority*, and you're supposed to perform whichever of the two operations comes first from left to right, in the direction you read. For instance, if, in an equation, a division sign comes before a multiplication sign, from left to right, you divide first and multiply next.

AS: Many often interpret this as "A is before S," (as they think M comes before D) but in fact they are of equal priority in the way M & D are. Addition or subtraction should only be performed after all other operations are completed. Then you perform either addition or subtraction, whichever comes first from left to right. If subtraction comes before addition, you would do the subtraction first, then the addition next.

If anything, consider PEMDAS a *loose reminder* of the *complete* Order of Operations, although if it were up to me, PEMDAS be thrown away completely and replaced with **GEMA**:

1. Groups (simplify, inner to outer)
2. Exponents or roots
3. Multiplication or division
4. Addition or subtraction

Actually, since:

- Roots are technically a form of exponents, when converted to rational exponents[*];
- Division is technically multiplication of a fraction; and
- Subtraction is technically addition of negative numbers,

it could even be condensed to just:

1. Groups,
2. Exponents,
3. Multiplication,
4. Addition

*Note: Roots are converted to "rational exponents" when the radical sign is removed and the root-number is moved to the denominator of the exponent.

The Unwritten 1

You must remember that "1" is often not (required to be) written or shown, but is still there. I give an example showing the "1" written as both the coefficient in front of x (could be any variable), the denominator of x, the power of x, and the denominator of the power 1 (of x). Moving right, I show it with the denominators removed, and then all the 1s removed, showing just x.

$$\frac{1x^{\frac{1}{1}}}{1} = 1x^1 = x$$

However fundamental this may seem, it is a concept students often question or forget, and for that matter, sometimes fail to implement when necessary. Here are reasons it is helpful to remember that "1" is *still there*:

- As a coefficient so its associated variable can be added to other like-terms, such as: $x + 3x = 1x + 3x = 4x$

  or as in adding radicals: $\sqrt{5} + 2\sqrt{5} = 1\sqrt{5} + 2\sqrt{5} = 3\sqrt{5}$

- As a denominator, especially for (fraction conversions to like-fractions during) addition/subtraction of fractions as in:

$$\frac{3}{5x} + 2x = \frac{3}{5x} + \frac{2x}{1} = \frac{3}{5x} + \frac{2x(5x)}{1(5x)} = \frac{3}{5x} + \frac{10x^2}{5x} = \frac{3 + 10x^2}{5x}$$

- As a denominator for (inverting, then multiplying a fraction during) division of fractions as in:

$$\frac{\frac{7x}{2}}{\frac{2}{3}} = \frac{\frac{7x}{1}}{\frac{2}{3}} = \left(\frac{7x}{1}\right)\left(\frac{3}{2}\right) = \frac{21x}{2}$$

- As a power or root, especially for multiplying factors (of a common base) with exponents (in which you add the exponents), as in:

$$x(x^3) = x^1(x^3) = x^{(1+3)} = x^4$$

- As a power or root, especially for dividing factors (of a common base) with exponents (in which you subtract the exponents), as when simplifying:

$$\frac{x}{x^9} = \frac{x^1}{x^9} = x^{1-9} = x^{-8} = \frac{1}{x^8}$$

- As a valid placeholder after a Greatest Common Factor has been factored out, as in: $3x + 3 = 3(x + 1)$

It is also worth reminding you about the unwritten one associated with a negative sign. Take a look at the following examples: -4, -x, -$\sqrt{1}$. These can be thought of as:

(-1)(4),   (-1)(x), and   (-1)($\sqrt{1}$), respectively.

Property Crises of Zeros, Ones & Negatives

There are many fundamental properties involving various operations with 0, 1 and negative numbers (I will focus mostly on "-1"). Some are easy to remember, however, some are easy to confuse or forget, but they are vital to get right. In textbooks, these are often thrown at you from different directions, at different times, often with vocabulary or definition-like labels. These are properties involving multiplication, division, exponents and roots. For ease and convenience, I've summarized the important ones here in this section, leaving out the labels, but showing the property, then explaining it in words, the way you might say it, hear it, or hear it in your head. Hearing (or reading how you might hear) these should drive home an extra dimension into your brain, so you can more easily recall them later.

$(\#)(1) = \#$.
In words: Any number times one equals itself, always, with no exceptions.

$\# \div 1 = \#$,   also seen as a fraction: $\frac{\#}{1} = \#$
In words: Any number divided by one equals itself, always, with no exceptions. For fractions: Any number over one equals itself (the top number). Likewise, any number can be assumed to be over one, and can easily be converted to a fraction by putting it over one.

Any # or term $\div$ itself $= 1$,   also seen as a fraction: $\frac{Any\ \#\ or\ term}{itself} = 1$.
In words: Any number or term divided by itself equals one, except when that number is zero. For fractions: Any number (or term) over itself equals one, except when those numbers are zero. Another way to say it is, "any non-zero number over itself equals one." See the next example.

12

$\# \div 0 =$ undefined, also seen as a fraction: $\frac{\#}{0} =$ undefined.

In words: Any number divided by zero is undefined. Or as a fraction, any number over zero is undefined. Any number divided by zero *does not equal* zero. You might say "you can't divide any number by zero."

$1 \div 0 =$ undefined, also seen as a fraction: $\frac{1}{0} =$ undefined $=$ undefined.

In words: One divided by zero is undefined, because any number divided by zero is undefined, as shown in the previous example.

$0 \div$ any $\# = 0$, with the exception of when the denominator is 0;

Also seen as the fraction: $\frac{0}{\text{any} \#} = 0$, except when the denominator is zero.

In words: Zero divided by any number equals zero, with one exception. The exception is: Zero divided by zero is undefined, because any number divided by zero is undefined. Another way to think of it is:

$0 \div$ any non-zero $\# = 0$, and

$$\frac{0}{\text{any non-zero} \#} = 0.$$

$0 \div 1 = 0$, also seen as a fraction: $\frac{0}{1} = 0$.

In words: Zero divided by one equals zero. In fraction form: zero over one equals zero.

This exemplifies two other properties, previously shown. This example follows that:

- Any number over one equals that number (the top number), and
- Zero over any non-zero number equals zero.

$0 \div 0 =$ undefined, also seen as a fraction: $\frac{0}{0} =$ undefined.

In words: Zero divided by zero is undefined, because any number divided by zero is undefined.

- This is the *exception* to the rule that "any number over itself equals one."
- It is also the *exception* to the rule that "zero divided by any number equals zero."
- Zero divided by zero *does not equal* zero, as some mistakenly think.

This property is especially useful when looking at:

- Slopes of Vertical Lines (see: A Vertical Line, and: When $x_1 = x_2$), and
- Extraneous Solutions (see: Solving Equations with Rational Expressions and Extraneous Solutions).

So far, we have looked at many properties and examples which result as: "Undefined". When doing these functions on a calculator, you might get "error." For a more on that, see: What Does "Error" Mean?

All but one of the properties shown so far involve division with 0 and 1. The following section will focus on exponents and roots involving 0, 1 and negative numbers.

$Base^1 = Base$

In words: Any base to the power of 1 = the base ...which is the same as to say:

$Any\#^1 = itself.$

In words: Any number (base) to the power of one equals itself (that base), as in the next example:

$1^{Any\ \#} = 1.$ In words: One to the power of any number equals one, with no exceptions.

$0^1 = 0.$ In words: Zero to the power of one equals zero.

$1^0 = 1.$ In words: One to the power or zero equals one.

$Any\#^0 = 1.$ In words: Any number to the power of zero equals one (with one exception; see next example). Another way to remember it is:

Any non-zero$\#^0 = 1.$ In words: Any non-zero number to the power of zero equals one.

0^0 = undefined (and *does not* = 1 or 0).

In words: Zero to the power of zero is undefined. Likewise, zero to the power of zero *does not equal* one or zero.

15

$\sqrt[root]{radicand}$ It is important to remember the names of the components of a radical. The term *inside* the radical is the *radicand.* The term in the "v" is the *root.* Also, although the symbol, shape and setup of a radical may closely resemble long division, they are not the same in any way.

$\sqrt[2]{1} = \sqrt{1} = 1$ In words: The square root of one is one. Note: The square root means "to the root of two," but the "2" is commonly unwritten. When the radical sign has no number written in the "v" area, it is implied to be two, meaning the square root.

$\sqrt[any\ non-zero\ \#]{1} = 1$
In words: Any non-zero number root of positive one equals one.

$\sqrt[0]{Any\ \#} = undefined.$ In words: Any radicand to the root of zero is undefined. The "root of zero" can also be called the "zeroth root" or the "zeroeth root." Also, sometimes this answer is given as "infinity (∞)," instead of undefined.

$\sqrt{0} = 0$ In words: The square root of zero equals zero.

$\sqrt[any\ non-zero\#]{0} = 0$ In words: Any non-zero number root of zero equals zero.

$\sqrt{-1}$ = no real solution... and may be expressed as "i". In words: The square root of negative one has no real solution. It is said to have no "real" solution, because the symbol (letter) "i" (for "imaginary"... as opposed to "real") can be used interchangeably with $\sqrt{-1}$. In that way, you can consider $\sqrt{-1}$ as having a solution, but since it is not a real number, it is said to be "no real solution." The symbol i is useful to manage multiple occurrences of the square root of negative one within an expression or equation.

16

The following examples demonstrate the importance of the second order of operations, as well as the complete and proper wording of that rule, mainly that roots are to be computed first before any other multiplication and division or addition and subtraction. Notice specifically that the root is (to be) computed first, and any factors or negative signs outside the radical are applied next.

$-\sqrt{1} = -1$ In words: Negative of the square root of positive one equals negative one. It is important to notice that the negative sign is outside the radical and is thus attributed to the result of the radical, which is calculated first, according to order of operations.

$^{even\ \#}\sqrt{-1}$ = no real solution
In words: Any even number root of negative one has no real solution… and may be expressed in terms of "i".

$-\sqrt{-1}$ = no real solution… and may be expressed as "-i".
In words: The negative of the square root of negative one has no real solution, and thus might be answered as "no real solution," or as "negative i." It is important not to mistakenly see this as "a negative times a negative equals a positive," because the even root of any negative number has no real solution, first and foremost, regardless of if it is multiplied by a negative outside the radical.

Remembering that the square root (or *any even root*) of any negative number can't be found (yields no real solution) is important to remember when:
- solving quadratic equations,
- using the quadratic formula, and
- understanding the graphs of parabolas.

$\sqrt[3]{-1} = -1$ In words: The cube root of negative one equals negative one.

$\sqrt[odd \#]{-1} = -1$ In words: An odd number root of negative one equals negative one, as in the last example.

$-\sqrt{1} = -1$ In words: The negative of the square root of positive one equals negative one. This follows the order of operations by taking the root first (of positive one), then attributing the negative sign from in front of the radical next, which is really multiplying (the result of the radical, which here, is 1) by negative one (the "1" outside the radical associated with the negative sign is unwritten).

The following examples involving exponents also demonstrate the importance of order of operations, and the prioritization of groups (here, parentheses) and exponents, which precede multiplication (or division and addition or subtraction). The placement of a negative sign with regards to the parentheses and the base number is very important.

$(-1)^2 = 1$ In words: The base negative one squared equals positive one. For this to be true, the negative must be associated to the base: one *inside* the parentheses.

$(-1)^{even \#} = 1$ In words: Negative one to the power of any even number equals positive one.

$(-1)^3 = -1$ In words: The base negative one cubed equals negative one.

$(-1)^{odd \#} = -1$ In words: The base negative one to the power of any odd number equals negative one.

$-(1)^2 = -1$ In words: A negative outside the parentheses times positive one squared equals negative one. Order of operations dictates that exponents must be performed first, and positive one squared equals positive one. Since the negative is outside the parentheses, it's like multiplying negative one times positive one (positive one being the result of one squared).

Finally...

$-1^2 = -1$ In words: Negative one squared equals negative one.

Order of operations plays a prominent role in this answer. Specifically, the exponent must be applied to the base of one first, and the negative is applied next. This example is often mistakenly answered "positive one," because some mistakenly see this as the square of negative one, or negative one times negative one. This would be different if the negative were inside the parentheses with the one, and the power of two was outside the parentheses, as shown four and five examples back.

This example not only shows the importance of the placement of signs and parentheses (or lack of), but it also shows the importance of how the math is heard or spoken. If you look back at how this was translated *in words*, you can't assume that the negative is in parentheses with the base "1". For that reason, when speaking an equation containing parentheses, you should be specific as to where they start, end, and what is inside them.

I started you off with some fundamental properties. This review will help you brush up on the little things many people commonly forget from their previous math experience. Having these reminders will give you an extra edge in solving problems (especially problems that *appear* to be complicated, but aren't, once you apply these equalities).
For more on radicals and commonly used roots, see the: Radicals, Roots & Powers section later in the book.

Integers & Whole Numbers

The words "integers" and "whole numbers" are commonly confused and misused among students. These words are used to properly communicate math, so it is important you use them properly, as well as understand when they are used.

Whole numbers are *positive*, non-decimal, non-fraction numbers. The main aspect students often forget about *whole numbers* is that they can only be positive. They can also be defined as "positive integers." One place you will see whole numbers used are in chemical formulas and chemical reaction equations. The subscripts in a chemical formula can only be whole numbers. And the coefficients in a balanced chemical reaction equation also can only be whole numbers.

Integers are also non-decimal, non-fraction numbers, but can be positive *and* negative. Any and all whole numbers are also integers (whole numbers are the positive integers).

The word "integers" is often seen and used during factoring of trinomials and quadratics. When factoring using the Trial & Error/Reverse FOIL Method, you are told to look at the "integer factors of the first (ax^2) and last (c) terms" (although, sometimes fractions and decimals are permitted too, where applicable, which is generally when the first (ax^2) and last (c) terms are fractions or decimals to begin with).

Also, the absolute value of any integer results in a whole number.

Prime Numbers

A prime number is a number which is greater than 1, and is not divisible by any number other than itself and 1 (without resulting in a non-integer number). Meaning, if you divide a prime number by anything other than itself or 1, the result will be a decimal (or equivalent fraction). The first thirty prime numbers are listed here as a reference, so if you ever need a place to quickly check a number, you can refer here. You can easily find more on the internet.

2, 3, 5, 7, 11, 13, 17, 19, 23, 29, 31, 37, 41, 43, 47, 53, 59, 61, 67, 71, 73, 79, 83, 89, 97, 101, 103, 107, 109, 113...

Why are these important to know? Because knowing whether a number is prime tells you whether you can proceed to factor it. If it is *not* prime, you *can* factor it. If it *is* prime, you *cannot* factor it. This is useful to know when you are factoring numbers (sometimes using a factor tree; see your textbook for more on factor trees); when doing factor trees, your goal is to factor all numbers into prime numbers. This may be used when finding an LCD or in the process of factoring trinomials into binomials. You also use factoring with radicals, but in those cases, you don't always need to factor to prime numbers. For more on that, see: Manipulating & Simplifying Radicals.

Polynomials can also be prime, which in their case means they can't be factored into polynomials any smaller than themselves. For an example of a prime polynomial, see: The Sum of Two Squares.

It is also worth noting that the opposite of a prime number is a *composite number*, which is a number *with* non-decimal factors other than itself and 1.

Is 51 a Prime Number?

The number 51 *is not* a prime number, but is often mistaken for being prime. I guess it just somehow looks prime... whatever that means. It's clearly not even (so it's not divisible by 2), and it's just a number we don't use as regularly as the numbers 0 through 50. Plus, many other prime numbers end in 1, as seen in the list of Prime Numbers. But don't let this number slip under your radar when you need to know whether it is prime.

How can you tell it isn't prime? Because it follows the technique that: if you add the digits of a number, and that sum is a multiple of 3, then that (original) number is also divisible by 3 (and another integer). The digits $5 + 1 = 6$, and 6 is clearly a multiple of 3, therefore 51 is also divisible by 3. The number 51 can be factored to (3)(17), which is actually the prime-factorization of 51.

These factoring strategies are discussed more in The Procedure for Prime Factoring. Also, you will notice the number 51 in The Prime Factor Multiples Table.

What is a Term?

A term is any number or variable or combination of them that either stands alone or is set apart by other terms with a "+" or "-" sign. Terms may be made up of factors, but terms are not factors themselves. They can't be factors because factors are multiplied, not split apart by plus or minus signs. Be careful not to use "factor" and "term" interchangeably.

What is a "Like-Term"?

A like-term is a term with a common base *and* common exponent. Terms must have *both* the same variable (letter) and the same exponent to be eligible to be combined. Only like-terms can be added or subtracted (*see note below). If there are terms that are not variables, they must only be constants (numbers) which count as like-terms and can simply be added or subtracted together.

"Like-terms" also apply to addition & subtraction of radicals. Since any radical or root can be converted to and expressed as an exponent (when it is, it is called a rational exponent), this follows the definition of "like-terms." "Like-radicals" can still be added and subtracted even if they're not in rational exponent form.

You may not deal with radicals when you begin the basics of algebra; you usually learn about these later. For more on how like-terms apply to radicals, see: Manipulating & Simplifying Radicals.

*Note: This does not mean like-terms cannot be multiplied or divided, because they can be. The statement referred to above is meant to accentuate that non-like-terms cannot be added or subtracted. To be clear, *any* terms can be multiplied and divided; they don't *have to be* "like." Non-like-terms *can be* multiplied (however, when they are multiplied, they are technically factors). Consider the terms: $3x^2$ and $4x^2$. They are "like" *and* can be multiplied to get $12x^4$.

What is a Factor?

A factor is a number or variable that is or can be multiplied by another number or variable. Factors combine via multiplication to make a term (and yes, factors are multiplied to give a *product*, but this section is meant to help distinguish between *factors* and *terms*, as they are often used incorrectly interchangeably). But often times, to serve their function, factors are not multiplied together; rather they are factored from larger numbers or terms and shown as unmultiplied, individual factors. Unmultiplied factors just look like numbers or variables standing next to each other (often each in parentheses). Converting a larger number or term into factors is done by factoring.

Factoring

Factoring is a way of breaking down a larger number or term into its factors. Factoring is performed using division and trial & error (explained in The Procedure for Prime Factoring). This is often done to compare factors to other factors, so when common factors are found, they can be cancelled out or grouped together by adjusting the exponent, depending on the situation. Also, terms are often factored into prime factors when you do a *factor tree*. For more on factoring numbers, see: Prime Numbers and The Prime Numbers Multiple Table.

Factoring terms with variables is a bit different. It is explained in The Greatest Common Factor, a few sections later.

Also, factoring polynomials is a bit different yet. Factoring polynomials in the form of trinomials and quadratics is briefly discussed in: Trinomials & Quadratics.

The Procedure for Prime Factoring

For smaller numbers, finding two starting factors may be easy to figure out. But for a large number, it may not be as easy, and this is a place student run into trouble. In this case, you can use a process which starts from an easy place, but you must know the process and follow it properly and sequentially. This process involves looking for a small prime number that the number you are factoring is divisible by (actually, you are looking for the *smallest* prime number that the number you are factoring is divisible by).

This is the process: Ask yourself if the number you are factoring is divisible first by the smallest prime number (2), then the next larger (3), and keep working your way up until you find it. You may need to check back to the list of Prime Numbers once you go beyond the prime numbers you know by heart. Within this process are sub-processes that you should apply as you work your way up the prime numbers. The following are those helpful sub-processes:

For 2: Check to see if the number you are factoring is even. Specifically, check to see if the last digit of the number is even.

     If it is even, then the original number you are factoring is divisible by 2. Then you should divide it by 2 to find the other factor.

     If it is not even, then the number is not divisible by 2, nor is it divisible by any other even number: 4, 8, 10, 12, etc. Next, test if it is divisible by 3...

For 3: Add up the digits in the number you are factoring. If the sum of the digits is divisible by 3, then the original number you are factoring is also divisible by 3. Then you should divide it by 3 to find the other factor.

     If the sum of the digits is not divisible by 3, then the original number is also not divisible by any other multiple of 3 such as 6, 9, 12, 15, 18, etc. Next, test if it is divisible by 5...

For 5: If the number you are factoring ends in either 5 or 0, then it is divisible by 5.

     If doesn't end in 5 or 0, then it is also not divisible by any other multiple of 5. Next, test if it is divisible by 7...

For 7 and higher: Take the number you are attempting to prime factor, divide it by 7, and analyze the quotient. If the quotient comes out as a whole number, then the prime number 7 you divided by is a factor, the quotient is the other factor, and therefore the number you are analyzing *is not* prime.

If the quotient comes out as a decimal, then the number you divided is not a factor. Repeat this process by attempting to divide by the next higher prime number until you get a quotient which is a whole number. You may need to repeat this process on the whole number quotient until you end up with all prime numbers (factors). The largest (prime) number to bother dividing your original number by can only be half the size of that original number.

For instance, if attempting to determine if the number 23 is prime by using this procedure, once you find that dividing by 11 results in a decimal number, dividing by 12 or 13 wouldn't matter at that point because there are simply no integer (non-decimal) factors greater than half of the number being analyzed.

You might also refer to my Table of Prime Number Multiples to expedite the process. Once you have found all the factors needed, or have completed the factor tree, present your factors according to your book or teacher's instructions.

The Prime Number Multiples Table

When students learn multiplication, they are sometimes given a table showing the numbers 1-10 (or higher) along the top (row) and down along the left (column), to easily find the product of two numbers. Students are also taught *prime numbers*, especially when learning to (prime) factor numbers. As in the example of "51" discussed previously, there tend to be some higher numbers which are not prime numbers, but may either appear to be, or students give up trying to figure out if they are through dividing up the prime numbers list, according to the suggested procedure for factoring large numbers (and finding what they are divided by). So I created a helpful tool I call a Prime Number Multiples Table. It is important you understand what it is and how to read it. This table is meant to be an extension for the procedure used to prime factor a number, according to The Procedure for Prime Factoring.

	3	7	11	13	17	19	23
3	9	21	33	39	51	57	69
7	21	49	77	91	119	133	161
11	33	77	121	143	187	209	253
13	39	91	143	169	221	247	299
17	51	119	187	221	289	323	391
19	57	133	209	247	323	361	437
23	69	161	253	299	391	437	529
29	87	203	319	377	493	551	667
31	93	217	341	403	527	589	713
37	111	259	407	481	629	703	851
41	123	287	451	533	697	779	943
43	129	301	473	559	731	817	989
47	141	329	517	611	799	893	1081
53	159	357	561	663	867	969	1173

First, you will notice that the numbers across the top row and down the left column are prime numbers only. Also, notice that the prime numbers 2 and 5 are not shown in the table. That is because it would be a waste of space because you can easily tell if any number, no matter how big, is divisible by 2 (because it would be even) or 5 (because it would end in 0 or 5). Even though the method for determining if a number is divisible by 3 is simple, I included 3 anyway, because you can't always tell if a number is a multiple of 3 just by looking at it, the way you can for an even number.

If and when you are attempting to factor a large number, you should look for that number in this table. If you find it, then you automatically know it is a product of the two prime numbers from the top row and left column it comes from.

If your number is:

- less than 1173,
- not found in this table,
- (and not a multiple of 2 or 5, as you should have checked in the beginning),

then your number *must be prime*. If your number is greater than 1173, then you must continue checking it (by dividing it) with prime numbers beyond the prime number 53.

For example, analyze the numbers: 209, 211, and 1193. First, look for them in the Prime Numbers Multiple Table.

- 209 is clearly in the table in 11's row and 19's column. Therefore it is *not* prime because it is clearly a product of numbers other than itself and 1, they being 19 & 11.

- 211 is not in the table and is clearly not even and does not end in 5. Therefore you can quickly conclude it *is prime*.

- 1193 is greater than 1173, the largest number in the provided table, so you must use the Procedure for Prime Factoring. Since 1193 is greater than 1173, is not even, and does not end in 5, you must divide by prime numbers to see whether it results in a non-decimal quotient. Since the largest multiple in the table is 53, you can begin dividing by 59. Dividing by all prime numbers up through and including 593 would result in a decimal. The next higher prime number is 599 which is greater than the half of 1173 (which is 596.5), so you can stop doing the division-by-prime numbers test. Therefore, 1173 is prime.

In short, this table will save you a lot of guess-work and trial & error. Although this is meant to be a time-saver and a tool to help you familiarize yourself with products of prime numbers, you still must be able to figure it out the long way (without the table), as I'm sure you won't be permitted to use this table or book during a test.

The Greatest Common Factor (GCF)

The GCF is *the largest* factor that can be factored out of every term involved. Another way to think of it is: the biggest factor that each term can be divided by, without resulting in a fraction or decimal. The GCF is found and used in the overall simplification process for:
- Reducing fractions, and/or
- Factoring a series of terms,

More specifically, the GCF is mainly found and used for two reasons:
1. To cancel common factors in a fraction to reduce that fraction. In this case, you would find the greatest factor that is *common* in the numerator and denominator and proceed to cancel them out (to 1).

2. To extract the GCF out of a series of terms, which, once you find it, you factor it out of each term, and the GCF goes outside (usually to the left of) a set of parentheses.

The Least Common Denominator (LCD)

The Least Common Denominator (LCD) is also known or used in some contexts as the Least Common Multiple (LCM). Actually, all LCDs are LCMs, but LCD is just specific to *denominators*. Books usually introduce the concept and procedure for finding the LCM in preparation for learning LCDs. In this book, I will refer solely to the LCD.

The LCD is commonly found and used for three main reasons:
1. To convert fractions into "like fractions" for adding and subtracting fractions (including: rational expressions), both of which are discussed later in the book.
2. To eliminate all denominators (and thereby all fractions) in an equation by multiplying each fraction by the LCD. In equations with fractions, this sometimes must be done in order to solve.
3. To reduce Complex Rational Expressions.

When doing problems involving an LCD, I recommend you write "LCD = (then show the factors here)" on your paper and fill in the factors as you gather them. This gives you a place of reference to keep track of your factors while you look back at the fractions in the problem. Sometimes you should use *factor trees* to help you find the LCD, especially if the LCD isn't obvious to you. Actually, using factor trees is a good habit. I often find that students want to avoid doing factor trees because they think the LCD is more obvious than it often is, but this is a major mistake. You have a higher chance of finding the correct LCD by doing factor trees. For more how to *find the LCD* and *factor trees*, please refer to your textbook, as I do not give the procedures in this book. But also, see: The Procedure for Prime Factoring.

GCF vs. LCD

Students commonly confuse the meanings and applications of the GCF and LCD, probably because they both involve factors and both are used for similar reasons. Both are used for simplification purposes and both can be applied in some way to fractions. In this section, I will go through a few brief hints to help you differentiate between the two.

First, you must understand their general uses, as you can read in the previous section. Next, let's break down the words and pay attention to some common, associated key words.

The best key words to associate with GCFs are: "out," "smaller," "found" and "division."
- GCFs are always factored *out* of a term or a number of terms, to make those terms *smaller*.
- When used with fractions, GCFs are factored *out* of the numerator and denominator to make fractions *smaller*. GCFs are factored out by *dividing* terms by the GCF.
- A GCF can only be *found*; it is not made. It is a factor that is *already there*, or *already within* the terms of interest. If there is no GCF present, then the fraction is already completely reduced, or the terms are already completely simplified.

- Whereas an LCD may already be there (if the largest denominator is *already* the LCD), but it can also be *made*...

The best key words to associate with LCDs are: "bigger," and sometimes "multiply," and "made." The starting point is looking at the largest denominator. The LCD used may be the largest of the original denominators (you would have to evaluate to make that realization). But if it is not, then:
- The LCD will ultimately be *bigger* than all the original denominators.
- The LCD will *made* by *multiplying* the appropriate factors.
- The fractions involved, which are to be converted (if necessary), will have their numerators and denominators *multiplied* to make fractions with *bigger* numbers than before.

These key words can seem a bit mixed when in reference to simplifying complex rational expressions. In this case, the LCD of all mini-fractions is either *already there*, or *made* (by *multiplication*), then *multiplied* by all mini-fractions involved. This ultimately *reduces* the complex rational expression, but along the way, some or all of the numerators of the mini-fractions become *bigger*.

Lastly, it is important how you use the word "find." In the GCF context, *"find* it", means *"look for* it." In the LCD sense, you *look* to see if the largest denominator is already the LCD, otherwise, you have to *find* (meaning *make)* it.

FRACTIONS

It is important you know how long division translates to fraction form, since fractions are a form of division, and you probably learned long division first. Also, you will revisit long division later when doing (long) division of polynomials (not covered in this book). It is also important to get the proper terminology down, which is often overlooked. An example of a long division setup is below, with the words in the proper places.

$$\text{divisor} \overline{)\,\text{dividend}}^{\,\text{quotient}}$$

The way you would say this is: "The dividend divided by the divisor equals the quotient."

Sometimes people use the word "in" when describing division. In that case, it would be
"How many times does the divisor go *into* the dividend?"
Answer: The divisor goes into the dividend (the) quotient number of times.

Consider the example:

$$2\overline{)6}^{\,3} \quad \text{which is the same as: } \frac{6}{2} = 3.$$

In this case, you might ask:
"How many times does two go into six?" In this case, you are saying:
"How many times does the divisor two go into the dividend six?"
Answer: Two goes into six 3 times. Which is the same as: "Six divided by two equals 3."

Notice how the divisor and dividend translate into a fraction and long division, and vice versa:

$$\frac{\text{dividend}}{\text{divisor}} = \frac{\text{numerator}}{\text{denominator}} = \text{quotient}$$

$$\text{denominator} \overline{\smash{)}\text{numerator}}^{\displaystyle \text{quotient}}$$

The dividend is the numerator. The divisor is the denominator. And the denominator goes into the numerator.

It is also worth noting that when performing long division (including synthetic division) of either numbers or polynomials, when there is a remainder, the remainder is properly reported by placing the "remaining part(s)" over the original divisor, as a fraction, listed at the end of your numbers or terms in the quotient.

Procedure for Adding & Subtracting Fractions

1) Find the LCD.

2) Convert all fractions to like-fractions (unless a fraction is already in correct form) by multiplying the numerator and denominator of each fraction by the missing factor which will make the current denominator the LCD.

3) To the right of the equal sign, write the LCD in the denominator and perform the operations (addition and subtraction) of the newly converted numerators from the left in the new numerator, over the LCD, on the right of the equal sign.

4) Simplify the fraction completely.

- Simplify the numerator, if possible, by combining like-terms, then
- factor if possible, and then
- reduce the fraction, if possible.

Multiplying Fractions

Start by attempting to cross-cancel common factors. Then, separately, multiply all the numerators together, and multiply the denominators together. Then evaluate what you have. It is often taught that fractions, once combined, should be expressed in lowest terms (also called reduced or simplified form). If you properly cross-cancelled before multiplying, your answer will come out in lowest terms. But sometimes students either forget to do this step, or just miss a set of common factors to cancel out. If you forget, that's ok, your answer won't come out wrong, but the numbers will be bigger and you will have to continue to factor. Evaluate your answer and look for a (greatest) common factor to cancel out at the end. To recap, you can either:

- Look for common factors in the numerator and denominator first, prior to multiplying, and cancel them out (this is called *cross cancelling*) – which results in a reduced, more manageable fraction, or you can

- Do the multiplication first, then factor the numerator and denominator, and cancel out common factors last.

The truth is, you can do it either way, and sometimes you just end up doing a mix of both to achieve the correct, reduced form of the fraction, which is fine and normal. But ideally, it is better to cross cancel first. For more on this, see: Factoring and: What is a Factor? Also, for more on cross cancelling, see: Cross Multiplication vs. Cross Cancelling.

Dividing Fractions

Dividing fractions is much different than multiplying fractions. Actually, only the first major step is different. Then it becomes multiplying fractions. In short, you multiply the first fraction times the reciprocals of the fractions being divided. But here is a more specific step-by-step procedure, followed by some comments.

1. Simplify numerators and denominators separately, and each fraction separately.
2. Keep the first (left-most) fraction the same (meaning: do not invert it... actually, you should re-write it *as is* on the next line down).
3. Invert (flip upside-down) all fractions that are to be divided (they will have a division sign to the left of them at first). Once fractions are inverted (they are now called *reciprocals*)...
4. Change what were division signs to multiplication signs (a dot, or put each fraction in parentheses). Many students overlook this simple step.
5. Now multiply all fractions, using the procedure for multiplying fractions.

Note 1: Just to be clear, when dividing a string of fractions, keep the first one on the left the same and flip each remaining fraction upside down, before multiplying. When a fraction is flipped upside down, it is called the *reciprocal*. It is also referred to as the "inverse fraction" or simply the "inverse.")

Note 2: Do not attempt to cross cancel factors before flipping the fractions and multiplying. Save cross cancelling until after the fractions are flipped and the division signs are changed to multiplication signs.

Note 3: If you attempt to divide numerators across the top and divide denominators across the bottom (in the way you would do when *multiplying* fractions), you will notice... it works! However... you are not encouraged to do it that way for one simple reason: it can get very complicated along the way, giving you strange fractions to manage, and many places to make a mistake. For this reason, you are highly encouraged to closely follow the procedure of flipping, then multiplying. It's easier, and if nothing else, it is much faster.

38

OPERATIONS OF BASES WITH EXPONENTS

Multiplying *Bases with* Exponents

When multiplying numbers or variables with a common base, keep the base the same and add the exponents together. Remember... when multiplying factors (with a common base) with exponents, you do not multiply their exponents; this is a frequently made mistake.

Dividing *Bases with* Exponents

When dividing numbers or variables with a common base, keep the base the same and subtract the exponent of the denominator from the exponent in the numerator. Remember to keep the signs of the exponents. You may subtract a negative exponent, yielding a positive exponent. You may also get a negative number as the exponent, which is fine, but in the final, simplified form of your answer, you shouldn't leave an exponent negative. If an exponent is negative, move the factor (the base) with that exponent to the opposite part of the fraction and change the sign of the exponent to positive.

Remember, any and every factor has an unwritten exponent of "1". Also remember that when exponents add or subtract to equal zero, any base to the power zero equals 1. (Review this in: The Unwritten 1, and: Property Crises of Zeros, Ones and Negatives).

Exponents of Exponents (a.k.a. Powers of Powers)

When you take a power to a power, multiply the exponents. Remember that if there are multiple factors, you must distribute the outer exponent to the exponent of each factor in the parentheses, including the coefficient. To distribute the outer exponent to each exponent of factors in the parentheses means you multiply those exponents. Remember, a variable with no exponent shown really is to the power of 1, and must not be forgotten to be multiplied by the outer exponent. This is a frequently made mistake.

Also, when distributing an exponent, do not forget to apply that power to the coefficient if there is one. This is another common mistake, often forgotten by students. This may be because students look for the conspicuous exponents written with the obvious variables, but when coefficients don't have exponents associated with them, they are just shown with an inconspicuous unwritten power of 1.

SOLVING SIMPLE ALGEBRAIC EQUATIONS

Solving a Simple Algebraic Equation with One Variable (First Degree)

The goal is to completely isolate the variable and to have it equal a number, which is the answer. Although this may seem easy (and it will become easier as you practice), it can also be complicated (if not just tedious), and students who are learning this for the first time often underestimate the importance of doing this in an orderly, systematic way. If you don't learn to do this properly, you will quickly get left behind in class.

This is done in two general steps:
1. Isolating the "term with the variable,"
   - Using addition and subtraction, and then
2. Isolating the variable
   - Using division or multiplication.

The following is a chronological list of detailed instructions to help you.

1. If there are any denominators, find the least common denominator and multiply all terms on both sides by the LCD to eliminate all denominators.

2. Simplify: Identify and combine like-terms, if any.

Note: #s 1 & 2 are interchangeable. You can simplify (combine like-terms) first and eliminate denominators next. However, it is usually easier to "get rid" of fractions *first*, so you don't have to go through Adding & Subtracting Fractions, which can just be more tedious.

3. Isolate the "term with the variable." Use addition or subtraction to "move*" the constants (non-variable numbers) to the right of the equal sign and...
- Use addition or subtraction to "move*" the term(s) containing variables to the left of the equal sign.
* I use "move" in a context which indicates the use of the addition principle of equality in which you add the opposite of the term you want to *move* (because adding opposites equal zero, *canceling* out a term), and what you add/subtract to one side of the equal sign, you must do to the other side to maintain the equality.

4. Simplify by combining like-terms. There should be one term (the term with the variable) on one side and a number on the other side.

5. Isolate the variable: Multiply both sides by the reciprocal of the coefficient in front of the variable. If the coefficient in front of the variable is negative, you should multiply both sides by the negative reciprocal in order to eliminate the negative sign and make your isolated variable positive.

You should now be left with a variable equaling a number.

Arrangement: Descending Order

It is always best to put all terms in descending order – from highest power to lowest power, from left to right. This organization facilitates easier simplification. Put all terms in descending order, even terms within parentheses and groups.

One reason (descending) order matters is for factoring. It is easiest to factor polynomials (like trinomials into binomials) when you see the terms in descending order. It will also help you identify and cancel out common factors (when they are polynomials) when the factors inside the parentheses are in descending order.

Another reason terms need to be in descending order is for long division of polynomials. Because of the systematic process of long division, the divisor and dividend must both be in descending order.

Expressions vs. Equations

It is very important to know the subtle difference between an *expression* and an *equation*. Simply put, expressions are not equations. Expressions are combinations of terms and operation symbols with no equal signs. Since expressions do not have equal signs, they cannot be solved, they can only be simplified. Equations are solved. Books often focus on expressions to stress and practice simplification. This is necessary (although sometimes misleading... I'll explain why shortly), because equations contain expressions.

Equations are mathematical sentences that contain an expression or expressions, and equals signs, and can be *solved*. The steps to solving an equation involve simplification of the expressions within the equation.

As I was saying above, books focus a great deal on expressions, which is fine, but this is why it can be misleading. The books go by a bottom-up approach and narrow focus on simplifying (factoring) expressions only, before incorporating those applications towards *solving* equations. What happens is: students get in the mindset of simplifying or factoring an expression (only), and stopping, that (later during solving) after simplifying, they forget to solve the rest, usually not more than two simple steps from the end-point. This is especially evident when students are supposed to solve quadratic equations. A common mistake is that students will successfully factor the trinomial in an equation but then forget to solve for the variables. So keep this goal in mind:

44

Factoring an expression is only *part* of solving an equation. Once you successfully factor an expression, get in the habit of continuing on to solve the equation. Those steps are discussed in the next section.

Author's Note: If it were up to me (and someday, I hope it is… I hope to write an entire algebra text book), the lessons on learning factoring *and* solving would be consolidated into one lesson, to better connect the reasons for learning factoring to solving equations and graphing. In the meantime, I hope this book helps you realize that these concepts are closely connected and not just separate entities.

Below are simple examples of an expression and an equation. Notice the small details which set them apart.

An expression: $3x^2 + x - 10$

An equation: $3x^2 + x - 10 = 0$, or

Also an equation: $y = 3x^2 + x - 10$

LINEAR EQUATIONS

A linear equation is an equation of the first degree; it produces a straight line. Lines are generally known to have:
- a slope (m),
- one y-intercept (b),
- one x-intercept (there is no symbol, but the x-intercept is x when y = 0), and
- the (slope-intercept) form: $y = mx + b$.
- It could also be in standard form.

However, there are circumstances in which they will not have all of these criteria. I will summarize these and the three types of straight lines next.

A Diagonal Line:

A diagonal line will have a y-intercept, an x-intercept, and a slope of anything other than zero (or undefined). It will be in the form of $y = mx + b$ (when it is in slope-intercept form).

A Horizontal Line:

A horizontal line will be in the form: "y = a number."

It will have a y-intercept, and more specifically, the equation will be:

y = the y-intercept.

For instance, if the y-intercept is 3, the equation of the line will be "$y = 3$".

Students also mistakenly think the equation for a horizontal line will be in the form "$x =$" (I think) because they associate the "x-axis," with "horizontal." But the opposite is the case, as explained above.

It is worth noting here that the equation for the x-axis is "$y = 0$" because it *intersects* the y-axis at "$y = 0$".

Also, a horizontal line *will* have a slope of *zero* and will *not* have an x-intercept.

Students often mistakenly say horizontal lines have "no slope" (because the slope is zero), but this is incorrect. "No slope" does not mean "zero". For more on this, see: The Slope Equation, and When $y_1 = y_2$.

A Vertical Line:

A vertical line is in the form of "x = a number."

A vertical line will have neither a y-intercept nor a slope, but it will have an x-intercept.

More specifically, the equation will appear in the form:

x = the x-intercept.

In other words, if the line intersects the x-axis at -5, then the equation for the line will be "x = -5".

Students sometimes mistakenly think the equation of a vertical line will be in the form
"y =" (I think) because they associate the y-axis with being vertical.

It is worth noting here that the equation of the y-axis is "x = 0" because it *intersects* the x-axis at "x = 0".

Another common mistake is to use "zero" and "no-slope" interchangeably, but they are significantly different. A vertical line literally has *no slope* (not even zero). It can also be said that the slope of a vertical line is "undefined." For more on this, see: When $x_1 = x_2$, and: What Does "Undefined" Mean?

The concept of horizontal and vertical lines (and their equations) is something that students often have trouble with at first, perhaps because the books seem to give them a small section displaced from the more emphasized diagonal lines (which is just inevitable). Nevertheless, a good way to gain a stronger understanding of horizontal and vertical lines is to graph them (that way you can see them, and we all know what horizontal or vertical look like), and to have a good understanding of calculating slope, as shown in: The Slope Equation.

What Does "Undefined" Mean?

There are a variety of circumstances where "undefined" is used to describe the outcome of an equation. "Undefined" can sometimes be used in a similar context as "No Solution," such as when computing an operation that can't be done, like dividing by zero. Sometimes "Undefined" is used for times that a computation can't be done, but isn't referring to the answer of a problem. This could be the case when looking at the slope of a vertical line. The slope isn't the "solution," so "No Solution" isn't appropriate... you would say the slope is undefined, or has "no slope."

The most important thing about understanding "Undefined" is not using it synonymously with "Zero." Also, when you do a computation on a calculator which would result as "Undefined," your calculator will show "Error."

How to Graph a Linear Equation

You can graph any equation… so don't be afraid to do it at any time! Making a graph, whether you are asked to or not, is a great way to give clarity to your problem or answer, and is especially a great way to help you understand a problem or equation from a more visual and conceptual perspective. Graphing a linear equation is the easiest of all the types of possible equations.

To make a line, you need two or three or more points. Two points are the minimum number of points needed to make a line, but having a third point is better. Having a third point is a good check mechanism because if the three points do not fall into the same line (and instead, make a triangle), you know at least one of the points is wrong, and you must go back and correct it. If this is the case, I recommend starting your table of points over, since you won't truly be able to tell which point (if not multiple points) is wrong. Also, the more points you have, the more accurate your line will be.

When dealing with linear equations, remember this:
When in doubt, make a graph.
To make a graph, make a table of 3 or more points.
Use the following procedure. First, draw the table, then fill in "0" for the first x, "0" for the second y, and "1" for the third x, as shown below.

x	y
0	
	0
1	

Then take each number, substitute it into the original equation and solve for the other variable. This will give you three important points:

    (x,y)
    (0,) ← the y-intercept, also known as b, or as a point (0, b)
    (,0) ← the x-intercept
    (1,) ← another easy point to find, near the origin

Sometimes, these points overlap, such as when the x-intercept and y-intercept are both at (0,0); or when the y-intercept is (1,0). That's fine. Just make another point on the table. My next choice would be to put in "1" for y, then solve for x. You can really choose any starting number for either x or y, then substitute it in and solve to find its counterpart variable.

Here is another related piece of advice: If your slope is a whole number, write it over 1. For instance, if your slope is found to be m = 3, write it as $\frac{3}{1}x$, because this will remind you that there is a rise *and* a run when you draw the graph.

If you have two equations (and their lines) to compare, be sure to make two separate tables so you can differentiate which points belong with which line. This could be useful for solving a system of two linear equations.

The Slope Equation

One major component of lines and graphing linear equations is the slope. The following shows all the interpretations of slope:

$$\text{slope} = m = \frac{\text{rise}}{\text{run}} = \frac{\text{vertical}}{\text{horizontal}} = \frac{\Delta y}{\Delta x} = \frac{y_2 - y_1}{x_2 - x_1} = m$$

The symbol Δ is the capital Greek letter "D" which stands for "the change in," commonly used in math and science equations.

The 4 Important Equations for Lines

These equations should be memorized, names included.

1. Slope-Intercept Form: $y = mx + b$

2. Standard Form: $ax + by = c$
   (See More on Standard Form on the next page)

3) Slope: $m = \dfrac{(y_2 - y_1)}{(x_2 - x_1)}$ from the points: (x_1, y_1) & (x_2, y_2)

4) Point-Slope Formula: $y - y_1 = m(x - x_1)$

Important comment about the Point-Slope Formula:
Keep y as y and x as x! Do not attempt to substitute values in for those here! You need them to remain (as letters) to the end of the process. The purpose of this formula is to substitute only values in for x_1 & y_1 (from a point) and the value of m, and then rearrange it into either Slope-Intercept Form or Standard Form.

Look at the name… it's the *POINT-SLOPE* formula… don't overlook the name! You need one (x,y) *point* and the *slope* to substitute into it, which can be rearranged into
$y = mx + b$.

Sometimes, you will be given 2 or more points and no slope (m) and will be asked to find the equation of a line (as $y = mx + b$). In this case, you must:

- First calculate m by using the two given points (or, if more than two are given, you must select any random set of two points) to put into the slope formula, and calculate m.

- Next, choose one of the given points and put the corresponding values in for y_1 and x_1 and m (that you just determined) into the *point-slope formula*.

- Then, use the proper methods (rules of equality) to rearrange the point-slope formula into the slope-intercept formula,
  $y = mx + b$.

52

More on Standard Form

Standard Form Linear Equations go by the following criteria:
- Terms with variables on the left of the = sign, in alphabetical order.
- c, representing a constant (a #), on the right of the = sign.
- The leading coefficient is positive.
- There are no fractions or decimals.
- The equation is in reduced form.
- a, b & c are #s, including possibly zero, but
  - both a & b can't be zero at the same time.
  - If $a = 0$, the line is horizontal.
  - If $b = 0$, the line is vertical.

Note: the "b" here is not the same "b" as the y-intercept in the slope-intercept form. Although the same letter is used in each, they are used in completely different contexts. The letters "a" and "b" are typically used to represent coefficients in front of variables.

Also Note: A Standard Form Linear Equation is slightly different than a Standard Form Quadratic Equation.

If the leading coefficient is negative, multiply every term by -1 to convert it to positive (and the signs of every term will change).

If any term contains a fraction, multiply each term by the LCD to remove all denominators.

If any term contains a decimal, multiply each term by the proper factor of 10 to remove the decimal(s).

When $x_1=x_2$:

- the slope is always undefined (and said to have "no slope"),
- the line is vertical, and
- the equation for the line will look like "x = #".

Consider this example of the equation of a line going through the following points: (4,3) and (4,7). Notice the x-values are the same, both 4, so in the equation:

$$\frac{y_2 - y_1}{4 - 4} = \frac{y_2 - y_1}{0}$$ ... the slope, m, is undefined because of the zero in

the denominator. The equation of the line here is "x = 4"

When $y_1=y_2$:

- the slope is always zero,
- the line is horizontal, and
- the equation of the line will look like "y = #"

Consider the following example of an equation of a line going through the two points (2,5) and (3,5). Notice the y-values are the same, both 5, and in the equation for slope:

$$\frac{5 - 5}{x_2 - x_1} = \frac{0}{x_2 - x_1}$$ ... since the numerator is zero, the slope, m, of the

line is zero, and the line is horizontal. The equation of this line would be "y = 5"

Parallel & Perpendicular Lines on a Graph

Lines are *parallel* when their slopes are identical.

In order to see this, you either need to:
- Rearrange both equations into slope-intercept form and look at m, or
- Simply calculate m for each equation and compare them.
- You can also get a good idea by graphing and looking. If the lines cross, it will be fairly obvious.

It is not recommended to evaluate the slope (m) when equations are in standard form (or any form other than slope-intercept form). For more on this, see: No Solution - Inconsistent.

Lines are *perpendicular* when their slopes are exactly both opposite *(and)* reciprocals of each other. For example, if one slope is 4, the other must be $\frac{-1}{4}$. For more on this, see: One Solution - Consistent.

SOLVING A SYSTEM OF (TWO) LINEAR EQUATIONS

Before continuing, there are some important things to be aware of.

A *system* of equations means: lines on the same graph that may intersect.

What does it mean to *solve* a system of two linear equations? To solve means to find the *point* of intersection, which is literally in the form (x, y)… so you're essentially finding an x and corresponding y. The (x,y)-point *is the solution* (when there is a solution, which there won't always be). For more on this, see: Interpreting the "Solutions", including: No Solution, in the next few pages.

You always need as many equations as you have unknown variables to solve for. Here, there are 2 equations and 2 unknown variables. (If you have 3 unknown variables, you would need 3 equations with those variables, etc.).

What Does "Solving In Terms Of" Mean?

When rearranging equations, you will often hear it explained as "Solve for one variable *in terms of* the other variable." More specifically, you might hear it as "Solve for y *in terms of* x." It is important that you understand the context of the words "in terms of." In the beginning of algebra when you learn to solve simple algebraic equations, you are used to solving one equation with one variable, and finding the numeric value of that variable; that is the endpoint. From that point, students often get accustomed to: solving, and getting a number for an answer, and being done.

But sometimes, especially for a multi-step problem (as a system of two linear equations is), you solve for (isolate) a variable, but you don't get a number-answer (at least right away). But this is ok. Some students who expect to solve and get a number (instantly) think they made a mistake when they don't get a number. This is not unusual. This is where one variable is solved *in terms of* another variable, meaning neither variable goes away, and you don't solve to get a number... but you are still rearranging the equation to *isolate* a variable of interest, and whatever variable(s) still remains is shown on the other side with the other terms. To *solve for* something *in terms of* a variable can be translated and broken down like this:

"Solve for" = isolate
"Something" = whatever you're isolating. It could be a variable. It could also be a number or an entire term. Whatever it is, get it to one side, and whatever remains goes on the other side.
"In terms of" = the variable(s) or term(s) which go on the opposite side of the entity being solved *for*.

Note: This type of equation rearrangement is not only used for solving systems of equations. You will often also see a small chapter in your textbooks dedicated just to multi-variable equation rearrangement. The equations used are often associated with geometry, trigonometry, statistics, economics, physics and chemistry. Exercises in solving multi-variable equations in terms of other variables prepare you for actual application of those equations in their related fields.

The Three Ways to Solve Systems of Two Linear Equations

There are three general ways to solve a system of two linear equations (two lines in a two dimensional space, also known as a plane):

1. Graph & Check
2. The Substitution Method
3. The Addition/Elimination Method

All three methods will yield the same outcome. I will discuss the best time to use each method, especially when it is better to use the Substitution Method vs. the Addition/Elimination Method.

How many solutions should you expect? There will either be:
- one solution (made up of one x and one y),
- no solution, or
- infinite solutions

These are the only possible outcomes. For instance, there can't be two solutions. This explanation is continued in: Interpreting the Solutions, in which the situations and associated vocabulary for these solutions are explained in more detail. But the next three sections give a closer look at each of the three solving methods.

Graph & Check

Sometimes this method is broken up separately into *Check* or *Graph*, and sometimes they must (or can) be done together.

Often the Check-only method is introduced first. In this case, an ordered pair (an x, y point) will be given to you. To check, you must substitute the x-value in for x and the y-value in for y into both equations, then simplify. Your goal is to determine if the point is or is not a solution. In order for the point to *be* a solution, *each equation* will reduce to *a number that equals itself*, such as $3 = 3$. But if only one (or neither) simplifies to *a number that equals itself*, then the proposed point is not a solution. You will know when the point is *not* a solution because an equation will simplify to a number that *appears* to equal a different number, which reveals itself to be a blatant *inequality*, and therefore not a solution.

In another related instance, the point itself isn't given to you, but a pre-drawn graph is. In this case, you are expected to *read* the point of intersection, and then *check* that point as explained in the previous paragraph. Often times, a pre-drawn graph will be made to have an obvious point, and by *obvious* I mean "integers," not some obscure decimal numbers.

Or you may have to do every step: create the graph, interpret the point of intersection and check it, as previously described. You could be given two sets of points (two or three sets of points for each of the two lines), which you will have to plot and graph. You might also just be given two equations and be instructed to "find the solution," in which case you must:

- Determine three points for each graph (so a total of 6 points),
- Plot the points and sketch the two lines,
- Make a judgment as to the point of intersection, then
- Check the point by substituting the values into each of the two equations, simplifying, and interpreting the outcome.

A few reminders are listed on the next page…

A few reminders on the Graph & Check Method:

- Use the procedure for: How to Graph a Linear Equation.

- Upon making a table of points to plot, take notice of any x-y points that are the same in both tables. If you find a matching pair, *that* is your solution. You should probably still graph, then check to confirm.

- Neatness is essential when graphing. Try to use graph paper. If you don't have any, consider using a straight edge. But more importantly, try your best to draw each unit on your x and y axes with equal lengths. This will make your point of intersection more accurate, and will give you a better ability to find the correct point, which should then allow it to successfully check.

- Also, even when you are not *required* to Graph & Check, you still can if you want to double check your results from the Substitution Method or the Addition/Elimination Method.

The Substitution Method

The substitution method is started in one of two ways.

One way is by taking one equation and solving it for one variable. When doing this, aim for the variable that will be most easily isolated. A good way to identify the best variable to isolate is by finding a term with either a small coefficient, or a term with a coefficient that the other terms in the equation will be easily divisible by. For instance, if one of the terms is "2y" and the other two terms are even (perhaps they are 6x and -8), then solving for y is a good choice because the other terms are (easily and noticeably) divisible by 2.

Sometimes a term already has no coefficient (meaning it has an unwritten coefficient of 1). In this case, it's a good idea to employ the Substitution Method because part of the work is done for you (that part being to make its coefficient 1). All you have to do then is isolate that variable. There are many systems of equations where this is the case. Often times, the writers of the math problems set you up to notice this variable. Sometimes, the variable is even already isolated, so all you have to do is realize that, then proceed to substitute what it equals in for that variable in the other equation.

Once you have solved for one of the variables *in terms of the other variable*, you must take that equality and substitute it into the variable for which you just isolated, in the *other* equation. For instance, if you just solved for y in one equation, then you must take what y equals and substitute that *in for y* in the other equation. When you do this, you should take notice of three things:
1. All the variables in the equation you just substituted *into* will be the same. It is only when they are the same that allows you to solve for the numeric value of that variable.
2. Once you do the substitution, you will be solving for the numeric value of the *other* variable. For instance, if you originally solve for y in terms of x and then substitute *in for* y in the other equation, you will then solve for the numeric value of x. Also,
3. Be careful not to make the common mistake of substituting into the equation you just solved for in the first step. If you do, then once you simplify, you will end up with a number that equals itself, and this may leave you confused and wondering where to go from there. So remember to substitute into the *other* equation.

You are now one step away from completing this problem, but students sometimes get confused at this final step.

The last step is to take the numeric value of the variable you just solved for and substitute that back in for that variable into *either* of the original equations, then solve for the other variable. For instance, if you just found the value of x to be -5, substitute -5 back in for x in one of the original equations, then solve for the value of y. I have two comments about this:

- Students are often confused by: Which of the original equations should I substitute my value back into? And the answer is: Either. Sometimes the choice seems to confuse students, so here's how you can choose. You can either just randomly pick one, or

- Choose the equation which appears to simplify easier. The one that will be easier to simplify may be the one with smaller coefficients or the one that does not contain fractions. If both look like similar difficulty, just choose one randomly. The answer will come out the same.

Sometimes students forget the final step, perhaps because, up to this point, you are used to a one-number or one-variable answer. Don't forget this step. Remember, the solution is a point (an x and a y).

The most common mistake made by students using this method is getting confused about what to substitute. So in summary, you solve for one variable *in terms of the other*… leaving you with one variable isolated on one side of the equal sign, and the other two terms on the other side. You then substitute *in for* the variable you just solved for into the *other* equation. The "stuff" you substitute *into* the *other* equation will replace the variable with the two terms from the other side of the first equation. You must then distribute and simplify in order to solve for your first value.

The Addition/Elimination Method

Some books call this "The Addition Method" and some books call this "The Elimination Method." I call it a hybrid of both, because you start by *adding* the two equations (after any necessary conversions), which *eliminates* one variable, making it a new, one-variable equation that can be solved (for the other variable). Remember, to perform this method, terms of the same variable in each equation must be *opposites* so they cancel out to zero when (the equations are) added together.

But my focus is to tell you *when* it is advantageous to use this method. Here are some clues to look for:

- You notice two terms of the same variable in each equation which are *already opposites* [meaning same term (variable and coefficient) but opposite signs]. These are already set up to cancel each other out to zero once the equations are added. All you have to do is add the equations, then proceed to the next step.

- You notice that one term is one multiple away from making it the *opposite* of a term of the same variable in the other equation. For instance, if one term in one equation is 3x, and the other equation has a -9x, then 3x can become +9x by multiplying it by 3 (and don't forget to multiply that factor through by the other terms in that equation). Or you have the term -5x in one equation and -5x in the other. You must multiply one of the equations through by -1, to make a -5x become +5x.

- Or, whenever neither equation is given with a variable with (an unwritten) coefficient of 1... mainly, the opposite of what is explained for The Substitution Method. In other words, all variables *have* coefficients (other than 1).

Common mistakes are described on the next page...

Common Mistakes:

- A very common mistake students make is adding the two equations *without* checking and converting one equation (to manipulate one variable into the opposite of a term from the other equation). If you do not properly set the equations up to have opposite terms, then adding the equations will just give you a third equation, still having two variables. Students often get stuck here, and rightfully so, because this is a dead-end; there's nothing you can do with it.

- Another common mistake students make is forgetting to add the constants when the equations are added. The constants are the numbers with no variables attached. Don't forget to add them, as they are just as much a part of the problem as the terms with variables.

Examples for Choosing the Method

In this section there are two "systems of linear equations" given: System A and System B. Each equation is already simplified and put into standard form for this type of problem. I want you to examine each set, and using the clues explained in the previous two sections, determine which method (Substitution or Addition/Elimination) would be best to use in each case. The answers and explanations will be given on the following page. The "solutions" to the system will also be given in case you want to do the problem for practice.

System A:
Equation 1: $3y + 5x = -13$
Equation 2: $y - \frac{1}{2}x = 0$

System B:
Equation 3: $3y + 14x = 1$
Equation 4: $-2y - 7x = 7$

System A would best be solved using the Substitution Method because the "y" in Equation 2 already has an (unwritten) coefficient of 1. The next step would be to isolate the y by adding $\frac{1}{2}x$ to both sides, giving you: $y = \frac{1}{2}x$. Then substitute the $\frac{1}{2}x$ in for y in Equation 1. The solution to System A is (-2, -1).

System B would best be solved using the Addition/Elimination Method. There are two ways to approach this. First, take notice of the "14x" and the "-7x". If the "-7x" is multiplied by "2," it will become "-14x" which is the opposite of "14x". You would need to multiply each term in Equation 4 by 2 to properly convert "-7x" into "-14x". Then, when you add the two equations, the "x" terms will cancel out to zero, allowing you to then simplify and solve for y (and then x). The solution to System B is (2, -14).

Or, you could solve System B another way, by multiplying Equation 3 by "2" and multiplying Equation 4 by "3". This would make
"3y" would become "6y,"
"-2y" would become "-6y," and
the y-terms cancel out to zero because 6y – 6y = 0.

As I mentioned before, some students take a mistaken approach to this first by adding the two equations together without multiplying through the necessary term(s) to manipulate terms of one variable to cancel. If mistakenly added, you would then get (what I will call)
Equation 5: y + 14x = 8

… Notice how both variables still remain in the equation? This leaves you at a dead end, because you can't successfully use this equation to solve for either variable.

Interpreting the "Solutions"

One Solution - Consistent
When the two lines cross, this is called a *consistent* system. In this case, there is one solution to be found, which is the point of intersection. The lines can be a combination of diagonal, horizontal and/or vertical lines. Keep in mind, all sets of *perpendicular lines* have one solution and make a consistent system.

No Solution - Inconsistent, Parallel
The two lines don't cross... because they are parallel; parallel lines by definition never touch. This is an *inconsistent* system. There are 3 ways you can tell that lines are parallel:

1. When using one of the three methods for solving a system of two linear equations, when you simplify and get to the end of the problem, you will get one # that does not equal the other #. It will look something like:

$$7 = -5 \quad \text{or} \quad 0 = 4, \quad \text{which clearly isn't true.}$$

2. The slopes (m) of the two lines are identical. In order for you to see this, you must convert the equations into slope-intercept form (y = mx + b), and then simply look at the slopes. The equations of lines may or may not originally be in slope-intercept form (y = mx + b). If they are not, convert them to slope-intercept form by solving for (isolating) y.

    Also, be sure each equation is in simplified form. If there is a Greatest Common Factor in an equation, you must factor it out. If you don't, the slopes may appear different, even though, by proportion, they are actually the same.

3. Graph and look. You can find out if lines are parallel without graphing, as described in the last paragraph. But this method (graphing & looking) should act as a backup to the two methods above, to confirm your answer. It may also be a great help if you are a more visual learner. For a reminder on graphing from an equation, see: How to Graph a Linear Equation. Once drawn, look at the lines to see if they appear to cross.

67

Infinite Solutions - Dependent

The entire lines overlap... because they are essentially the same line. The graph actually looks like one line. This is called a *dependent* system. There are 3 ways to tell this:

1. When you use the Substitution Method for solving a system of two linear equations, the equation you substituted *into* will simplify to something like this:

$$4 = 4 \qquad \text{or} \qquad -8 = -8 \qquad \text{or} \qquad 0 = 0.$$

It won't even let you get to the point where a variable equals a number, revealing that the system is *dependent*.

Note: When this happens, students tend to think they made a mistake because this outcome seems so awkward, but usually they haven't made a mistake... it's supposed to turn out this way to indicate that it's a dependent system.

2. When reduced to simplest terms and converted to same form, the equations are identical. If you are going to compare equations, they must both be in the same form as each other (either slope-intercept form or standard form).

Note: Often times, these equations may look similar before they are completely simplified. If they are in the same form, you may notice the coefficients are different, yet proportional. This can be a sign that dividing one or both equations through by a certain factor will then reveal the equations to be identical.

This is why it is so important to try to simplify any equation by looking to factor out a GCF and arranging into standard form in the beginning of every problem. Doing this here would instantly reveal that the system is dependent.

3. Graph & Check – Graph both lines and look at the graph. It should be pretty obvious that the lines overlap. Actually, it will just look like one line.

TRINOMIALS & QUADRATICS

The words "trinomials" and "quadratics" are often used interchangeably because they overlap, both in characteristics, looks and application (particularly during factoring and solving). Despite their similarities, they should not be seen as completely synonymous by definition. Because of the way many books and lessons are arranged, sometimes these are seen and used too disconnectedly or separately. This is understandable as well (when done correctly), however this may also mislead students to miss the important connection and overlap between them. This section is to help you clearly relate and differentiate the similarities and differences between them, by definition and use.

A *trinomial* is an expression containing three different terms, often with at least one squared variable. I say "often" because when you are introduced to factoring (using the Trial & Error or Reverse-FOIL method, for instance), you are factoring trinomials into binomials. By definition, trinomials can be comprised of *any* three terms to any power, but trinomials are very often directly associated with factoring into two binomials as the segue-way to solving *quadratic* equations.

A *trinomial* sometimes overlaps as a *quadratic* expression and may be part of a *quadratic* equation. Although an expression may be both a trinomial and a quadratic expression, they are not synonymous by definition. A trinomial is a quadratic expression when the highest power (degree) of any term is 2. When they do overlap, they can be simplified (factored) the exact same way. Also, not all quadratic expressions are trinomials, as you will read next.

A *quadratic* equation is:

- An equation containing a squared variable (like x^2), yielding a maximum of two solutions [but could contain one solution (that occurs twice), or no solution].
- It must contain a squared variable and thus is considered a 2nd degree equation.
- Although a quadratic equation can contain a term of x (to the unwritten power of 1), it can never contain a term of a power higher than 2.
- Also, a quadratic equation, when graphed, always makes a parabola (a U-shaped curve).

A *quadratic* equation appears in the standard form:

$$ax^2 + bx + c = 0$$

There are a few things you should understand about the equation written above.

It may also be written as:

$$y = ax^2 + bx + c = 0, \text{ in which } y = 0, \text{ as above, or:}$$

$$f(x) = ax^2 + bx + c = 0,$$

because quadratics (which make parabolas) are considered to be "functions." Specifically, they are functions of x.

(I do not delve into "functions" in this book, but if you're wondering, an equation is considered to be a "function" if its graph crosses the y-axis once and only once.)

It is written in descending order and standard form in this case. For a quadratic equation, "standard form" means all terms are on one side of the equal sign, and set equal to zero (on the other side). *Descending order* means the terms are arranged from the highest to the lowest power, from left to right.

All quadratics are not always originally presented in descending order or standard form. If and when they are *not*, you should rearrange each one into standard form and descending order before simplifying and solving.

70

The letters a, b and c are representative of numbers, not variables (more on that down the page). Also, "a" cannot be zero. If "a" is zero, it is no longer a quadratic equation; it is then a linear equation.

A quadratic equation contains a trinomial expression when a, b and c are all non-zero numbers. However, sometimes, either the coefficient b, or constant c, or both, are zero (remember, "a" cannot be zero). This is worth highlighting because this is where students often run into trouble. I believe they run into trouble at first because, when b or c is zero, the equations just look differently, and usually the solving method is different. For that reason, there a segment dedicated to those specific cases. I will show what they look like, explain their graphical significance, how to solve them, and their expected solutions. This is continued in: Quadratics With Zero. But first you should understand the solutions to quadratic equations.

What Are "Solutions" to Quadratic Equations?

It is good you know what solutions to quadratic equations are, to give you a better, overall perspective. (The variable of a quadratic equation is usually x, but can be other letters). When the variable is x, the solutions are "x-intercepts," which are simply the points on a graph where the parabola crosses (or the single point which touches) the x-axis; the x-intercepts are defined the same, no matter what type of equation or graph they come from. And x-intercepts are points (ordered pairs) at which $y = 0$, which is why you set your quadratic equation equal to zero at first (in other words, making sure it is in standard form). This is also why, if you solve by factoring, you set each factor equal to zero; or, if you solve by the quadratic formula, this is why the formula is set equal to zero. You will learn about this more in the next section.

All quadratic equations produce a parabola when graphed. The solutions are the x-intercepts of the graph. As you will see in the next few sections, there are a number of ways to solve quadratic equations. If you solve by factoring, you will get one or two solutions; those solutions will be either integers or fractions. If you solve by the quadratic formula, your answers may come out to be integers, fractions or radicals (which could be converted to decimals for graphing).

Also, you may find that a quadratic equation has either one, two, or "no real" solutions. Here is a quick summary of each scenario:

- *One solution* means that the parabola only touches the x-axis once; it does not cross the x-axis. You may think of it as "sitting on" the x-axis. Another more technical way to say it is, "The x-axis is tangent to the vertex of the parabola."

   This will occur when the trinomial factors into a binomial squared. It is also good to know that a binomial squared comes from a "perfect square trinomial."

- *Two solutions* means the parabola crosses the x-axis twice.

- Finally, you may find *"no real solutions"*. This means that the parabola does not cross or touch the x-axis, but don't be fooled. Just because it doesn't cross or touch the x-axis doesn't mean it doesn't exist... it still exists, and can still be graphed. This conclusion can only be made through use of the quadratic formula. If a quadratic equation is prime (which only means it can't be factored), this is still not grounds for saying "no solution"... it may just mean the solutions are radicals. But it may also mean there are "no real" solutions. The reason this is specifically answered as "no real" solution, instead of no solution, is because this is often the result of the square root of a negative number. For more on that, see: The Square Root of Negative One, and: Prime vs. No Solution.

These concepts and a closer look at the solving methods behind them are discussed in more detail in the next few sections.

Solving Quadratic Equations

Trinomials and quadratic equations can be solved in three general ways:

1. Factor & Solve
1b. Take the Square Root of Both Sides
2. Use the Quadratic Formula
3. Graph & Check

When using factoring to solve, there are actually three different factoring methods you can use, so you might say there are five or six total possible ways to solve quadratic equations, (although, as you will see, the factoring method(s) won't always work).

You also see "Taking the Square Root of Both Sides" in the list as "1b." This is an alternative method to certain cases in which factoring can be used as well. There is a type of equation which can be solved in two ways, either by "Factor & Solve" or by "Taking the Square Root of Both Sides". This is explained in: When both b & c are 0: $ax2 = 0$.

I re-wrote the list of ways to solve quadratic equations below, with the more specific, sub-methods included, so you get a concise list of methods and choices you can use.

1. Factor & Solve:
   - Trial & Error/Reverse FOIL Method
   - The ac/Grouping Method
   - The Complete the Square Method
1b. Take the Square Root of Both Sides
2. Use the Quadratic Formula
3. Graph & Check

In the following sections, I will go over the *when* as opposed to the *how* (see your text book for the "how"). I have good reasons for this. The textbooks usually do a good job of showing you how to implement the methods, and the steps are not really that complicated; a lot of practice is the key to becoming good at factoring and solving quadratic equations. However, the books don't usually answer a question many students have, which is, "when is the best time to use each method?" I'm going to answer that question, as well as give more of a top-down perspective on solving quadratic equations.

When dealing with quadratics, you should also get accustomed to starting them (or preparing them) the same way, no matter which method you use to solve them. You should always:
   - Look for a GCF to factor out (this is a big one students often forget to do), and
   - Arrange into descending order and standard form.

Factor & Solve

Usually, you should try to factor and solve a quadratic equation (before using the quadratic formula) because it's faster and involves fewer steps (if it's able to be factored). Factoring and solving can lead you to the answer(s), however if you can't factor, this leaves your answer inconclusive. If a quadratic is prime (can't be factored), it doesn't necessarily mean there is no solution, but you must then use the quadratic formula to come to that conclusion (to either find the answers or find that there is no real solution).

But many quadratic equations *can be* factored. There are three general ways to factor, but more importantly, there are better times to use each method and clues to dictate when those times are. Although I do not teach you how to do each method (as I stated, your textbooks do a good job at that), I will highlight the clues and tell you the best time to use each method.

Before factoring, you must go through a series of steps to set up and prepare your equation, no matter which method of factoring you will use. These are very important, and students often forget one or all of these because you don't always have to do them:

- Simplify as much as you can by combining like-terms, if necessary.
- Arrange all terms into Descending Order according as: $ax^2 + bx + c = 0$.
- Put into Standard Form by moving all terms to one side (the left) and setting them equal to zero.
- Look for a Greatest Common Factor. Sometimes you can factor and solve successfully if you forget to do this, but it will often leave all numbers larger, and the problem more tedious. If the GCF is a number, you can factor it out, then remove it (because if you divide both sides by it, the zero divided by it on the other side eliminates it, and the zero remains zero).
  - However, if the GCF is a variable or a power of "x", it won't be eliminated, but it will *equal* zero as one of your solutions. Factoring out a variable may allow you to properly factor using one of the factoring methods (including to factor again) that you otherwise wouldn't be able to do. Actually, when the GCF contains a variable or power of x, the problem might not have been a *quadratic* to begin with. Factoring that out may leave you with a quadratic that you can then factor by quadratic methods.

- Make sure the coefficient of the leading term (the "a" connected to x^2) is positive. You can't factor if it's negative (it can be negative, though, when using the quadratic formula). If it is negative, treat -1 as a GCF of each term. By factoring it out, you will simply change the sign of each term, and the zero on the other side isn't affected.
- (Optional) You may choose to eliminate all fractions first, if there are fractions. You are usually taught that it is a good rule of thumb to begin any type of problem by removing fractions by multiplying by the LCD. You don't have to though, and sometimes you can even factor them into binomials, but you will account for them in the final solving steps if you don't remove them first.

Trial & Error/Reverse FOIL Method

Various books have different ways of naming factoring methods. I use both these names here because I think they accurately describe the process they're used for. To do this method, you must simplify (combine like-terms), arrange all terms into standard form (move all terms onto one side), and put into descending order. You then pay attention to the factors of the first and last terms of the trinomial, write out the two binomials (or, once you get good enough at it, keep them in your head) then FOIL these factors to see if you arrive back at the original trinomial. If it works, you've found your factors; then set each binomial equal to zero and solve. Keep trying factors until you find the ones that work (that's what's *trial and error* about it). Also, I call it "Reverse-FOIL" because you are starting with the product (the original trinomial), then coming up with factors to *try*, then FOILing them to check. But as I stated, my intention is not to focus on the *how*, but the *when*. So when is the best time to us this method?

It is common to first approach a trinomial/quadratic expression with this method because, if it can be done (easily), it can be the quickest method with fewest steps. An equation with an expression of this type has the best likelihood to be solved by starting with the Trial & Error Method when "a" and "c" are relatively small. I use the word "small" loosely, and everyone's interpretation of "small numbers" may be a little different. My use of "small number" could be taken two ways. It could mean a number with few factors, or it could be taken more literally, meaning between 1 and about 15. Either way, the smaller the number, the fewer factors it will tend to have.

Ideally, prime numbers or numbers with few factors will yield the fewest possible combinations of factors. The fewer factor combinations there are, the fewer the choices there are to try (multiply and test). There's not really much else to say about this method. As the "a" and "c" numbers approach larger values, there could be so many possible combinations of factors to trial that it can become very time consuming. When the size and possibilities of factors seems overwhelming, this is a good time to do the "ac/Grouping Method."

The ac/Grouping Method

You will often first be exposed to the "Grouping Method" when you are learning factoring (before learning to solve quadratic equations and equations with trinomials). The Grouping Method is simply a method of factoring that is introduced to teach you how to factor four terms [if there are enough similarities (common factors) among pairs of terms]. Books don't often call the "ac Method" the "ac/Grouping Method;" they usually call it one or the other. This ac/Grouping Method is comprised of two main parts:

- First using the a and c to find the correct factors (by multiplying a and c, then looking at all the two-factor combinations of that product, called "ac"), then
- Writing out the four associated terms and using the *grouping method* to factor (into two binomials),

then solving.

When is the best time to use this method? It is worth noting that there are often two groups of students: those who prefer the Trial & Error/Reverse FOIL Method, and those who prefer this ac/Grouping Method. The reason I'm telling you this is because you can always skip the Trial & Error Method and start by using this method every time (in other words, you don't have to try the Trial & Error Method first, then go on to this method next if Trial & Error doesn't work out) if you like this method better. Some students prefer this method because it can be quicker on average, because it removes a lot of guess work (trial and error) and the time spent on all the *error* factor combinations.

Regardless, a good time to use this method is when the a and c factors are large and/or have many factors. To give you an idea of what might be "large" or "having many factors," if $a = 18$ and $c = -24$, there could be many factor combinations from them. The "18" has three pairs of factors $\{(1 \cdot 18), (2 \cdot 9), (3 \cdot 6)\}$, and the "-24" has eight pairs of factors $\{(-1 \cdot 24), (1 \cdot -24), (-2 \cdot 12), (2 \cdot -12), (-3 \cdot 8), (3 \cdot -8), (-4 \cdot 6), (4 \cdot -6)\}$. The negative sign doubles the number of factor combinations, because either factor could be negative.

Start by multiplying a and c; this gives you the product "ac," (it will be an actual number). You are then to look at every possible two-integer-factor combination of the product ac. You may consider setting this up in the following way: make two columns: one with the heading "factors," and the other "b." The point is to find the numbers that when multiplied give you the product of "a" times "c", and when added give you the middle term of the original trinomial, "b." When the "a" and "c" numbers of a trinomial are large or have many factor possibilities, this method will help you quickly find the combination needed to complete the "factoring by grouping" method. Again, don't forget to solve your binomials once you factor into them.

Complete the Square

Completing the Square is definitely in a category of its own. You may not even consider it factoring by the same definition as the other factoring methods, since it is more-so like a manipulation technique, involving factoring as a step.

This method is mainly used when conventional factoring doesn't work, because the
c-number may not factor into integers, but it can still be manipulated.

It's important to remember that to begin, the leading coefficient must always be positive 1, so before proceeding, if the coefficient is anything other than 1, divide each term by the "a" coefficient, and this will make the coefficient of the leading term "1" (and the other terms will change proportionally). Also, as you make your new "c-number," don't forget to add it to the other side, to maintain the equality. Students often forget to do the two things mentioned in this paragraph.

This creates a "perfect square trinomial" (on one side). A perfect square trinomial is a *special case*, in which the coefficient of the leading term will be an unwritten "1," (which is a perfect square), the new c-number you made will be a perfect square number [this new c-number may also be referred to as $\left(\frac{b}{2}\right)^2$, which is the formula for how to make it], and the coefficient b will be exactly two times the square root of the new c-number.

According to this special case, "A perfect square trinomial factors to a binomial squared."

But this slightly deviates from a regular problem where you're given a perfect square trinomial to solve. A perfect square trinomial will factor into a binomial squared, and when set equal to zero and solved will give you one answer. This is because a perfect square trinomial makes a parabola which touches (but doesn't cross) the x-axis, thus it has one x-intercept (solution). However, when you complete the square, your (what becomes a) binomial squared equals a number on the other side, and once it is solved, will result in two answers (not one).

The Quadratic Formula

The Quadratic Formula is:

$$x = \frac{\left(-b \pm \sqrt{b^2 - 4ac}\right)}{2a}$$ in which "a," "b" and "c" refer to the numbers from the

standard form equation: $ax^2 + bx + c = 0$.

The thing about this formula/method is that it always works. It will work when a quadratic equation can or can't be factored. Even if there is no solution, the endpoint of this method will reveal that.

About half the time, your solution(s) from using the quadratic equation won't be integers or even rational numbers. In those cases, the best way to express your answers will be in radical form (as opposed to decimal form). If you ever wonder why simplifying radicals is drilled into your minds, it's so you can use and navigate through the Quadratic Formula from beginning to end. But getting to the *very end*... the very last step is where students commonly make a mistake. Simply put, they often forget to finish it.

Note: "Standard form" is slightly different for quadratic equations than linear equations.

The Part Everyone Forgets (The Last Step of the Quadratic Equation)

Sometimes, at this point, your answer is completely simplified... but sometimes it's not. You should never assume it is completely simplified until you attempt this last step. Consider you've gone through the Quadratic Formula and get to this last step:

$$= \frac{9 - 3\sqrt{5}}{6}$$

Look for a GCF in the numerator (and factor it out), and factor the denominator. In this example, the GCF in the numerator is 3, which should be factored out. In the denominator, 6 factors into 3 and 2. Now it can be seen that the numerator and denominator have a common factor of 3:

$$\frac{3(3 - \sqrt{5})}{(3)(2)} = \frac{\cancel{3}(3 - \sqrt{5})}{\cancel{(3)}(2)} = \frac{3 - \sqrt{5}}{2}$$

Cancel the common factor of 3 out of the numerator and denominator, then re-write the simplified answer. If there is a radical in your answer and you must graph it (remember, an answer is an x-intercept point), convert your radical to a decimal and reduce everything to one number. Also, take extra care not to do the last step improperly, as many often do, as explained in: The Wrong Way To Simplify a Rational Expression

Graph & Check

Graph & Check is much different for quadratic equations than it is for linear equations. For example, Graph & Check for linear equations is used to find a solution of a system of two linear equations (the point where *two* lines cross). The circumstances (solutions) are different for quadratic equations because quadratics make parabolas. Therefore the way to find points to be *graphed* is different, because you can't just find and graph any three random points for a quadratic with a guarantee that they will be representative of the complete (parabolic) shape, as you would for a linear equation.

Also, the context of the word "solutions" is different for linear (systems) and quadratic equations. *Solutions* to quadratic equations are x-intercepts. [Just to be clear, you can (and do) find the x-intercept of a linear equation (you just don't call it the "solution"), and you could also plot two parabolas on the same graph and find their points of possible intersection… but again, those points aren't the primary contextual use of "solutions," and it's something you aren't commonly asked to do].

Here are the minimal points you need to graph a quadratic equation:
- The vertex
- The x-intercept(s), AKA the "solutions"
- The y-intercept
- Any additional points

Let's look at these points in greater depth.

Regarding the y-intercept, every quadratic (parabola) has one. If you have the equation written in descending order and standard form, it's the number which is "c" from
$ax^2 + bx + c = 0$. Every parabola will cross the y-axis (once), regardless of if it crosses the x-axis.

The vertex must always be found, as this is the (either maximum or minimum) point where the graph shifts from the positive to negative direction (or vice versa). This is its *inflection point*.

As for the x-intercepts, you may find:

- two x-intercepts, if the parabola crosses the x-axis,

- one x-intercept, if the (vertex of the) parabola touches but doesn't cross the x-axis (as is the case for perfect square trinomials), or

- no x-intercepts, if the parabola neither touches nor crosses the x-axis. For this, you will get "no solution", but the parabola may still exist.

This is where "any additional points" comes in. If you found the vertex, the y-intercept, and the x-intercepts, you can successfully sketch a decent representation of the parabola, by graphing the points and drawing the line through the points in a smooth, curved way (not in a rigid way as if you were playing "connect the dots"). Finding more points will just help you make a more accurate and complete curve.

Quadratics with Zero

This section is dedicated to showing you that the following cases *are* still considered quadratics (and therefore also second degree equations. Actually, quadratics in which b or c are zero are called "incomplete quadratics"). The coefficient "a" *must* be a number other than zero, otherwise, the equation would no longer be a quadratic. Also, keep in mind that when b = 0, the "bx" term equals zero and will not be written. Likewise, when c = 0, it won't be written. However, a zero *will* appear (only) if it is alone on one side of the = sign. Even though one or two terms within a quadratic can be zero, you may have to solve them differently than if all terms are non-zeros. Here, we look at those cases.

When c is 0: $ax^2 + bx = 0$

When $c = 0$,

$$ax^2 + bx + 0 = 0$$

which will be shown as: $ax^2 + bx = 0$

A few examples are: A) $4x^2 + 2x = 0$,

                                B) $3x^2 + x = 0$,

                                C) $-x^2 + 5x = 0$, or

                                D) $7x^2 - 3x = 0$.

In either case, you will always expect two solutions: $x = 0$, and $x = $ another #.

Note: In cases where "c is 0," then "$x = 0$" is *always* one of the solutions. From a graphical perspective, anytime c is 0 in a quadratic equation, the resulting parabola will always cross through the origin $(0, 0)$, as well as another point along the x-axis. As in any quadratic equation, "c" represents the y-intercept, which in this case is 0. In this case, the origin $(0, 0)$ is both the y-intercept and one of the x-intercepts.

These are the steps to solving:
- Factor out the GCF… which will include "x" and possibly a number as well.
- Set the outside factor(s) equal to zero, and in this case, this "x" automatically equals zero.
- Set what is inside the parentheses equal to 0 and solve for x.

Using the first example: $4x^2 + 2x = 0$,

factor out x and 2 (the GCF is 2x) from both terms:

$$2x(2x + 1) = 0,$$

set the outside factors equal to zero:

$$2x = 0,$$

divide both sides by 2, and thus

$$x = 0.$$

Set what's inside the parentheses equal to zero and solve for x:

$$(2x + 1) \rightarrow 2x + 1 = 0.$$

Subtract 1 from both sides: $\quad 2x + 1 - 1 = 0 - 1$

Giving: $\quad 2x = -1.$

Divide both sides by the coefficient 2, and $x = \frac{-1}{2}$

The solutions are $x = 0$ and $x = \frac{-1}{2}$.

When Both b & c are 0: $ax^2 = 0$

When both b and c = 0,

$$ax^2 + 0x + 0 = 0,$$

which will be shown as: $\quad ax^2 = 0.$

Some examples are: $\quad 2x^2 = 0, \qquad x^2 = 0, \text{ or} \qquad -3x^2 = 0$

In all cases, the only solution is x = 0 (because if you divide both sides by the coefficient in front of x^2, then take the square root of both sides, you will get "0").

Graphically, this will produce a parabola whose vertex is the origin (0, 0), with the line "x = 0" as the vertical line of symmetry. There is only one solution because (the vertex of) the parabola of this type touches but does not cross the x-axis. Here, the origin (0, 0) is both the y-intercept and (the only) x-intercept.

When b is 0: $ax^2 + c = 0$

When only $b = 0$,

$$ax^2 + 0x + c = 0$$

which will be shown and seen as: $ax^2 + c = 0$

Some examples are:

E) $2x^2 - 2 = 0$,

F) $9x^2 - 4 = 0$,

G) $x^2 - 36 = 0$,

H) $4x^2 + 25 = 0$, or

I) $3x^2 - 5 = 0$.

Of the given examples, all except "$4x^2 + 25 = 0$" are considered to be: "The *difference* of two squares." Example H is discussed a few pages later.

"The Difference of Two Squares"

We will look at each example, specifically, but before that, it's important you see the *two ways* in which problems like these can be solved, so you can notice the pattern in the examples to follow. Either approach is started in the same way.

First, look for a GCF. If there is a GCF, factor it out, then proceed to divide both sides by it. Since zero is on the right, any (non-zero) number you divide by it will equal zero. At this point, there are two ways you can proceed to solve.

One way is by moving the constant to the other side of the equal sign, then taking the square root of both sides. The other is by factoring into binomials. We will look at each method in more detail. Either choice is completely valid. It's really up to you to decide which method you prefer. To reiterate, the two ways are:

A1. Factor into conjugate pair binomials, or
B1. Take the square root of both sides.

There is a time that "taking the square root of both sides" will be preferable, as I will show in the following examples.

Example E: $$2x^2 - 2 = 0$$

This is a classic example of factoring out the GCF first, which here is 2.

$$2(x^2 - 1) = 0$$

Divide both sides by 2, which gives you

$$(x^2 - 1) = 0$$

At this point, you can go forward with either method. I'm going to demonstrate both, to prove that each is valid and yields the same outcome. But first, I'm going to show "factoring into conjugate pair binomials."

As "$x^2 - 1$" is the difference of two squares, it can easily be factored into

$$(x - 1)(x + 1) = 0$$

Set each factor in parentheses equal to zero, and solve for x.

$x - 1 = 0$ $x + 1 = 0$

$x - 1 + 1 = 0 + 1$ $x + 1 - 1 = 0 - 1$

$x = 1$ $x = -1$

so x = +/- 1

There is a very important lesson in this example, which is that

"the difference of two squares can be factored into *conjugate pair binomials*."

It should also be expected that "the difference of two squares" will always yield two opposite solutions (however there is one technical exception to this, explained in: Clarification: When the Solution is 0).

Conjugate Pair Binomials

It's good to be familiar with *conjugate pair binomials*, visually, by definition, by name, and by common use.

Conjugate pair binomials are the result of factoring "the difference of two squares."

They are known as conjugate pair binomials because...
- they are *conjugate* in that they are joined and connected in some way,
- they come in *pairs* (as conjugates do), and
- they are binomials (each set of parentheses contains two terms).

They appear as two parentheses with the same first and second terms, but opposite signs in between them.

When conjugate pairs are multiplied, the result is "the difference of two squares."

The advantage of identifying and factoring the difference of two squares into conjugate pair binomials is that it is quick and involves few steps.

Also the use of conjugate pair binomials are an essential part of "rationalizing the denominator" when the denominator contains a binomial with at least one radical. This topic is one that is usually covered near the end of the semester, often displaced from the lessons which introduce factoring quadratic equations and "special cases." During this time displacement, students sometimes forget to see the connection of this concept. Additionally, upon learning it, students often do not realize the relevance for which it will be needed later.

More on the procedure for multiplying conjugate pair binomials and graph-related information is covered in the Special Case subsection: *The Difference of Two Squares.*

The other way to solve for x, starting from "$x^2 - 1$" is by *taking the square root of both sides...*

Taking the Square Root of Both Sides

We start back with "$x^2 - 1$" from example E to see that it can also be solved by *taking the square root of both sides*. It is set equal to zero.

$$x^2 - 1 = 0$$

In this case, we proceed by moving the constant (here, -1) to the other side:

$$x^2 - 1 + 1 = 0 + 1, \text{ making it } x^2 = 1.$$

Take the square root of both sides:

$$\sqrt{x^2} = \sqrt{1}$$

Remember that taking the square root of a number gives the positive *and* negative root number (because if you square a positive or negative number, you always get a positive result), so it can be said that x equals plus or minus 1, written as

$$x = +/- \ 1.$$

The next two examples demonstrate "*factoring* the difference of two squares into conjugate pair binomials."

Example F: $\qquad$ $9x^2 - 4 = 0$

Factored: $\qquad$ $(3x - 2)(3x + 2) = 0$

Notice that 3x is the square root of $9x^2$ and 2 is the square root of 4. Set each set of parentheses equal to zero and solve for x.

$3x - 2 = 0$	$3x + 2 = 0$
$3x - 2 + 2 = 0 + 2$	$3x + 2 - 2 = 0 - 2$
$3x = 2$	$3x = -2$
divide both sides by 3	divide both sides by 3

$$x = \frac{2}{3} \text{ and } \frac{-2}{3}$$

Example G: $x^2 - 36 = 0$

Factored: $(x - 6)(x + 6) = 0$

$x - 6 = 0$ $x + 6 = 0$
$x - 6 + 6 = 0 + 6$ $x + 6 - 6 = 0 - 6$

          $x = 6$ and -6

Example I: $3x^2 - 5 = 0$

This is an interesting example, one of which you are sure to encounter. It's important to remember that you can take the square root of *any* (positive) number or term, but only when you take the square root of a *perfect square* will your result be integers (non-radical or non-decimal numbers). In the three examples before this one, the terms were *perfect squares* and therefore could be factored (into binomials), however in this example, the 3 and the 5 are not perfect squares (however, the x^2 still is)... so they *can't be factored* using integers. Therefore, when one or both of the terms involved are not perfect squares, it is often preferred to approach solving by moving the constant to the other side of the equation and taking the square root of both sides, as seen in the following steps.

$$3x^2 - 5 + 5 = 0 + 5$$

$$3x^2 = 5$$

Divide both sides by 3, giving:

$$x^2 = \frac{5}{3},$$

take the square root of both sides:

$$\sqrt{x^2} = \sqrt{\frac{5}{3}}$$

and

$$x = +/- \sqrt{\frac{5}{3}}$$

The Sum of Two Squares

In any quadratic equation in which b = 0, any time the c-number is
added, the polynomial in the equation is considered *prime*, and has no
real solution. (Please see the two notes below).

You should recognize equations such as these as:

"the sum of two squares,"

and you should remember:

"The sum of two squares is prime."

When faced with a problem such as this, the acceptable answer responses
are:
"prime," and "no real number solutions."

*Please Note: If you look at the standard form of a quadratic equation:
$ax^2 + bx + c = 0$, or in this case: $ax^2 + c = 0$, you probably notice that c is
added, which may lead you to think that every equation such as this is
prime, but this is not enough information to make this judgement. To be
clear, the form $ax^2 + c = 0$ is correct, but whether the equation is prime
all depends on the sign of the actual number that is represented by c. In
other words, if the number plugged-in for c *is negative*, you have the
difference of two squares, which *is not* prime. Or, if the number for c *is
positive* you have the sum of two squares, which *is* prime. There is
technically one exception to this, explained in: Clarification: When the
Solution is 0.

**Also Please Note: The first sentence of this section states, "... the
equation is considered prime, and has no solution." This statement must
be properly understood. Although the statement may seem to insinuate
that "prime" is synonymous with "no solution," by definition, this is not
true. Since the "sum of two squares" is so common, you can predict its
outcome of "prime" and "no solution" automatically, but the fact that
they have both these outcomes is merely coincidental. (This explanation
is continued in: Prime vs. No Solution). With that in mind, remember
that "the sum of two squares" can still be graphed and will produce a
parabola.

Now let's look at two more examples, J and K. Notice each example is similar to an example from before, but they both have a negative sign in the leading coefficient. What we're going to do is simplify each example first, and then decide if what remains is prime or factorable. To simplify, we must do a number of things:

- Identify and factor out a GCF (if there is one)
  - Divide both sides the by the GCF (if there is one)
- Move the constant to the other side of the equal sign
- Ensure the coefficient in front of x^2 is positive 1
  - This may have been taken care of in a previous step, otherwise...
  - You can do it now by dividing both sides by the coefficient in front of x^2

Look at Example J: $\qquad -4x^2 + 25 = 0$

Let's simplify by going through the procedure just previously mentioned:

- Is there a GCF? No.
- ...So there is no GCF to divide both sides by.
- Move the constant, 25, to the other side by subtracting it from both sides:

$$-4x^2 + 25 - 25 = 0 - 25$$

$$-4x^2 = -25$$

- Ensure the coefficient in front of x^2 is positive 1. In this case, we will divide both sides by -4, which cancels out both negative signs, giving:

$$x^2 = \frac{25}{4}$$

Take the square root of both sides... which we can do, because the sign of $\frac{25}{4}$ is positive.

The solutions are: $x = +/- \frac{5}{2}$.

Next, let's look at Example K: $-3x^2 - 5 = 0$

And follow the steps to simplify, as we did in the last example.

- Is there a GCF? In this case, yes. It is "-1"; factor that out:

$$-(3x^2 + 5) = 0$$

- Divide both sides by the GCF "-1", which makes it:

$$3x^2 + 5 = 0$$

If we pause here, we see that we have the sum of two squares, which is enough information for us to stop solving, and answer with: "prime; no real-number solutions"

Note: If you're wondering if 3 and 5 are square numbers, you could say "sort-of" because they are the squares of the square root of 3 and the square root of 5, respectively; however 3 and 5 are not *perfect* squares. Any positive number is a square of another number, but a *perfect square* is the result of squaring an integer. Let's proceed with the remaining two steps anyway to see what happens if we continue to simplify and solve.

- Move the constant 5 to the other side by subtracting 5 from both sides, giving

$$3x^2 = -5$$

- Divide both sides by 3 to ensure the coefficient in front of x^2 is 3, giving

$$x^2 = -5$$

At this point, when you go to take the square root of both sides, you should realize that you can't take the square root of a negative number and get a real-number solution. This further proves that "$-3x^2 - 5$" has no (real) solution.

In the previous few pages, we looked at a variety of examples of quadratic equations that appear differently due to various parts (b, c, or both) being zero, which are essentially *special cases*, and the corresponding ways to solve them. The next section looks at special cases from a different perspective, focusing more on visual "what to look for," and "what will it look like" clues and characteristics. They are also described *in words*.

Special Words for Special Cases

The textbooks are generally good at highlighting the special case (quadratics) in their own section, and teaching *how* to solve them. And students are generally good at factoring and solving them when they are in their own isolated areas and when they know what type they're dealing with. But once the *special cases* are mixed into general types of problems, students sometimes forget the signals in identifying them.

Identifying them is the first crucial step. This section contains a few sentences and key words that will help you identify the type of special case, and tell you what outcome to expect in terms of factoring, graphing, and the shortcut for multiplying. If you memorize these words, it will help you figure out the factors and answers more quickly. Some parts may seem redundant from some earlier parts of the book, but that's okay, the repetition is good for you.

I cover two main special cases in the coming pages. Please note that there is another common special case involving "the sum and difference of two cubes," which I don't cover in this book (because my primary focus is linear and quadratic equations, and squares and square roots).

Perfect Square Trinomial

Remember that: "A *perfect square trinomial* factors into a *binomial squared*."

Likewise, "A *binomial squared*, when multiplied out, gives a *perfect square trinomial*."

You will notice in a perfect square trinomial that:
- the first term (ax^2) and the last term (c) are perfect squares,
- the coefficient b will be the *product* of
  - (the square root of the first term, ax^2),
  - (the square root of the last term, c), and
  - 2;
- the sign in front of c will always be positive.
  - It is always positive because the last number in the binomial is squared.

It will resemble:

(Perfect square number)x^2 + $(\sqrt{a})(\sqrt{c})(2)x$ + perfect square number

Such as: $$4x^2 - 12x + 9 = (2x - 3)^2$$

ex.

with the perfect square trinomial on the left, equal to its factored binomial squared on the right; and another example:

$$x^2 + 8x + 16 = (x + 4)^2$$

There are two ways to multiply a *binomial squared*...

There are two ways to multiply a *binomial squared*. One way is by expanding it into two binomials and multiplying by the FOIL method. This way is fine, but you are highly encouraged to use the other way, which is the special case shortcut.

This is the shortcut for multiplying a binomial squared, in words. I will refer to the terms of the binomial as such: (first term + last term).

1. Square the first term.
2. Leave space (for what will be instruction #4).
3. Square the last term.
4. Multiply the first term times the last term, then double it, and write it in the space you left in instruction #2.

* A common mistake made is people treat this method like the shortcut for multiplying *conjugate pair binomials*, and they forget to do step 4, so they are then missing the middle (bx) term.

Do you see the connection of how the procedure for this special case is applied when completing the square?

When graphed, a perfect square trinomial is a parabola which touches (but does not cross) the x-axis. The vertex is at $y = 0$ and the solution to x, and (said to be) *tangent* to the x-axis.

The Difference of Two Squares

Remember that: "The *difference of two squares* factors into *conjugate pair binomials*."

Likewise, "*Conjugate pair binomials*, when multiplied, give the *difference of two squares*."

The difference of two squares is a specific case of when b is 0; $ax^2 + c = 0$, and "c" is a negative number. When "c" is a positive number, you have "the sum of two squares."
Remember: "The sum of two squares is prime."

There are two ways to multiply *conjugate pair binomials*. You can multiply them using the FOIL method, but you are encouraged to use the special case shortcut method. Getting used to the shortcut will be useful when you learn to rationalize denominators with radicals and binomials (not covered in this book). Plus, the shortcut is faster. This is the shortcut method in words:

1. Square the first term,
2. Write a minus sign,
3. Square the last term.

There will be no middle (bx) term because if you FOIL the conjugate pair binomials, the product of the Outer terms plus the product of the Inner terms cancel each other out to zero. This is what makes the *difference of two squares* a special case.

When graphed, this makes a parabola that:
- Has two x-intercepts because it
- crosses the x-axis in two places,
- has a vertex at $x = 0$, [at point (x, c)], and
- the line of symmetry is the y-axis (the line: $x = 0$).

There is one technical exception to the statement that "the difference of two squares yields two solutions," which is explained in: Clarification: When the Solution is 0.

Prime vs. No Solution

As in the section: The Sum of Two Squares, when the quadratic expression [in an equation in which b (only) is zero] is prime, the result is "no (real) solution." By definition, "prime" should not be thought as synonymous with "no solution." The following instances can occur:

- the polynomial *can* be factored and *does* yield (real) solutions;
- the polynomial *can't* be factored and yields *no* (real) solutions; and
- the polynomial *can't* be factored but *does* yield (real) solutions; however,
- a polynomial which can be factored will always yield one or more (real) solutions.

Keep their definitions in mind. "Prime" means "can't be further factored (into factors other than itself and 1)." And "No (Real) Solution" with regards to a polynomial means that the resulting graph has no x-intercepts (does not cross or touch the x-axis).

A good example is when you have a quadratic equation containing a trinomial expression, in other words, there are non-zero numbers in for a, b and c. However, as discussed earlier in: Solving Quadratic Equations, the trinomial may not be able to be factored using any combination of integer factors in the Trial & Error/Reverse FOIL Method, or the ac/Grouping Method; in this case, it would be considered "prime." You are then to use the Quadratic Formula. You may not yet know if the Quadratic Formula will yield "real" solutions or not (but that's why you plug in the a, b and c values and solve). If the Quadratic Formula *does* yield real solutions, the solutions will contain radicals of not-perfect squares (which could be converted to decimals). There is also a chance that *no* real solutions will result from that original prime polynomial due to a negative number in the simplified radical.

Clarification: When the Solution is 0

Before, I stated that *the difference of two squares* will yield two solutions, but there is one exception to this. Any quadratic equation in which b and c are zero, as seen in "When Both b & c are 0," can also technically qualify as the *difference of two squares*, such as "$4x^2 - 0 = 0$", or the *sum of two squares*, as in "$9x^2 + 0 = 0$", because zero is a (perfect) square number. In cases like this,

- there is *not two solutions*,
- nor is it *prime with no solutions*...
- It *has only one solution*, that being "$x = 0$".

That also makes this an exception to the statement: "The sum of two squares is prime and yields no real number solutions."

RATIONAL EXPRESSIONS

By its technical definition, a *rational expression* is a fraction that contains polynomials. But since, to be a polynomial, it must contain at least one variable (otherwise, if it just contained constants or numbers, it would be considered a regular fraction), my definition is: A *rational expression* is a fraction containing (one or more) variables. Notice the root word "ratio."

Sometimes, you will be asked to simplify a rational expression and sometimes you will have to solve an equation containing rational expressions. When dealing with rational expressions, you must know how to properly simplify them.

Procedure for Simplifying Rational Expressions

It is important and helpful that you can clearly see each numerator and denominator as separate pieces because they will need to be simplified individually, first, before continuing.

0. For this reason, I recommend putting parentheses around numerators and denominators. (This reinforces the first rule of Order of Operations, being that numerators and denominators are in fact "groups" but are rarely written with parentheses. Putting parentheses around them makes them look more like groups, and will remind you to treat them as such).

1. Factor each numerator and denominator separately, and completely factor the polynomials.

2. Look for factors that are alike in the numerator and denominator, then cancel them out to 1 (correctly, by avoiding: The Wrong Way to Simplify a Rational Expression).

106

Procedure for Adding & Subtracting Rational Expressions

This follows the procedure for Adding & Subtracting Fractions, only now, variables are involved. You add & subtract rational expressions the same way as fractions by finding and using the LCD. Remember, the goal for adding & subtracting rational expressions is to properly combine them so that one rational expression remains. Follow this procedure (the first steps, 0a – 0c, are more preparatory steps):

0a. Put parentheses around all numerators & denominators.

0b. Look for any minus signs in front of a fraction. If there is a minus sign in front of a fraction, you must distribute it through its associated numerator. To do this, replace the minus with a plus and change *every* sign in its associated numerator. This is often an overlooked step which results in +/- sign errors later on.

*Note: A big error students make here is applying the negative sign to only the first term in the numerator, instead of applying (distributing) it to *every* factor in the numerator.

0c. Put all terms in numerators & denominators in descending order.

1. Factor all the numerators & denominators separately.
2. Look at all factors of the denominators and determine the LCD.
   - Write the LCD off to the side so you can refer to it, and write the LCD in the denominator to the right of the = sign (as of now, the numerator is blank; you will fill in the numerator in step 5). Leave the LCD as *un-multiplied* factors (which will make simplifying easier in the later steps).
3. Look at each denominator (of each fraction) and determine what factors are missing (I call these the "missing-factors") if any, to complete the LCD in each.
4. Multiply the numerator & denominator of each fraction by its missing-factor(s)
5. Multiply & distribute factors in the numerator and write the products in the numerator above the LCD written on the right side of the = sign. Don't forget to transfer the proper signs.
6. Simplify the terms in the new numerator.
6a. Combine like-terms.
6b. Look for a GCF, and whether there is one or not, try to factor completely.

7. Simplify: Cancel out any common factors in the numerator & denominator.

- Make sure you avoid the wrong way to simplify a rational expression.

Simplifying a Complex Rational Expression

Simplifying a complex rational expression is something you will surely be required to do on a final exam.

A *complex fraction* simply means fractions within a fraction. Since *rational expressions* are fractions containing variables, a *complex rational expression* means rational expressions within rational expressions... or simply:
fractions-containing-variables within fractions-containing-variables.

There are two ways to simplify complex rational expressions:

1. The "All-LCD Method." Multiply all mini-fractions by the LCD of all mini-fractions.

Or

2. Simplify the overall numerator & overall denominator first (by applying the rules for addition & subtraction of fractions), separately, into one fraction each, then divide the top-fraction by the bottom-fraction (see: Dividing Fractions).

Either method, if used correctly, will yield the same result.

When should you use each method?

Method 1, the "All-LCD-Method" avoids needing to add & subtract fractions and you only need to find one LCD. This is typically easier and more preferred. When in doubt, default to using this method. Although this method is more straight forward, it is also tedious and may make your paper messy. For that reason, mistakes are often made in the process for not being able to read your own writing, writing too small, or not leaving enough room. When doing these problems by this method, be sure to leave plenty of room on the paper and write clearly.

Method 2, "Simplifying the Overall Numerator and Overall Denominator Separately" is typically used when the addition & subtraction of fractions in the overall numerator and overall denominator will be a quick and easy procedure. This method is also usually selected when the variables in the overall numerator are distinctly different than those in the overall denominator. The reason for this is because finding and using the All-LCD-Method may introduce variables into opposite parts of the fraction that will require extra and tedious steps (such as factoring) to get to the end.

Ultimately, the choice of method is more dependent on the student's preference, as both methods are tedious with a number of intermediate steps, but will still yield the same outcome. Because of how tedious they are, you may choose to stick with the one you gravitate towards and work on getting good at it. The two procedures will be given next, at first in a very detailed form, and then in a condensed form, so you can refer back to either version. These may help you decide on your preference.

All-LCD Method (detailed version):

0a. If there are whole numbers or polynomials that are not fractions, it's a good idea to put them over 1, to make them fractions.

0b. I recommend putting all polynomial numerators & denominators in parentheses. Books often leave them without parentheses, but using them makes it easier to view and use the polynomials or factors to continue.

1. Completely factor all numerators and denominators, if possible.

2. Find the LCD of all the mini-fractions involved.

3. Multiply the LCD by the numerators of all mini-fractions. (See Note 4, shown three pages later).

4. Simplify all mini-fractions by applying the rules of Multiplication & Division of Bases with Exponents. In this step, all denominators of mini-fractions should cancel out. The factors of the LCD are intended to directly cancel out (with) every entire denominator of all mini-fractions. However, this may still leave the remaining, un-cancelled factors from the LCD, if any, in the numerators of the mini-fractions, and this is expected.

The rational expression is now *simple*, not *complex*, as there is now only one numerator and one denominator.

     4a. Simplify the (new) numerator. Multiply, distribute and combine like-terms, where possible.
     4b. Simplify the (new) denominator. Multiply, distribute and combine like-terms, where possible.
     4c. Arrange all terms into descending order.

5. Completely factor the numerator and denominator, separately. That means factoring out a GCF first, if there is one, and/or factoring the polynomial, if possible, (into smaller polynomials).

6. Cancel out any common factors in the numerator & denominator.

*Note 1: Sometimes factors will cancel out in the last step and sometimes none will. Be ready for either scenario.

**Note 2: At this point, it is up to your professor if he/she wants you to multiply the factors in the numerator & denominator (individually) for your final reported answer. I personally prefer them to be left in factored form.

***Note 3: Don't commit the frequently made mistake: The Wrong Way to Simplify a Rational Expression.

(Note 4 is shown two pages later).

Simplify Overall Numerator & Overall Denominator Separately Method (detailed version)

0a. If there are whole numbers or polynomials that are not fractions, set them over 1, to make them fractions.
0b. I recommend putting all polynomial numerators & denominators in parentheses. This makes it easier to view & use these polynomials or factors to continue.

1. Completely factor all numerators and denominators, if possible.

2. Next, you will be simplifying the overall numerator & overall denominator separately, using their own LCDs.
2a. Find the LCD of the fractions in the overall-numerator
2b. Find the LCD of the fractions in the overall-denominator.

3. In the overall numerator and denominator, separately, use the Procedure for Adding & Subtracting Rational Expressions.

At this point, you still have a complex rational expression, but now with only one (unsimplified) mini-fraction in each the overall-numerator and overall-denominator.

4a. Leave all denominators of mini-fractions as un-multiplied factors.
4b. Simplify all numerators of mini-fractions by combining like-terms.
4c. Factor the numerators of the two mini-fractions, separately.

By this point, there should still be one fraction each in the overall-numerator and the overall-denominator, but now, each mini-fraction is simplified. You are now ready to...

5. Divide the top mini-fraction by the bottom mini-fraction using the rule for dividing fractions.

You now have a (simple) rational expression (one numerator and one denominator). They may or may not already be simplified.

6. Simplify: Cancel out common factors from the numerator & denominator.

All-LCD Method (short version)

0. Put whole factors over 1 and put all overall numerators & denominators in parentheses.
1. Factor all numerators & denominators.
2. Find the LCD of all mini-fractions.
3. Multiply all mini-fractions by the LCD.
4. Simplify: cancel out all denominators of mini-fractions with associated common factors.
*The rational expression is now *simple.*
4a. Multiply, distribute, combine like-terms in numerator.
4b. Multiply, distribute, combine like-terms in denominator.
4c. Put all terms in descending order.
5. Completely factor the numerator and denominator, separately.
6. Cancel out any common factors from numerator & denominator.

Simplify Overall Numerator & Overall Denominator Separately Method (short version)

0. Put whole factors over 1 and put all overall numerators & denominators in parentheses.
1. Factor all numerators & denominators.
2a. Find the LCD of the fractions in the overall-numerator.
2b. Find the LCD of the fractions in the overall-denominator.
3a. Convert and add fractions in overall-numerator.
3b. Convert and add fractions in overall-denominator.
4a. Leave all denominators of mini-fractions as un-multiplied factors.
4b. Simplify all numerators of mini-fractions.
4c. Factor both numerators of the two remaining mini-fractions.
5. Divide the top fraction by the bottom fraction.
6. Simplify the one remaining fraction.

Note 4: Using the LCD is similar, yet distinctly different in the All-LCD-Method than for *addition/subtraction of rational expressions* in the Overall-Numerator & Overall-Denominator Method. In addition/subtraction of rational expressions, the *missing factors* of the LCD are multiplied times the numerator *and* denominator of each fraction to convert the fractions into like-fractions, whereas for simplification of complex rational expressions, the whole LCD is multiplied times the *numerator only* of each mini-fraction.

Annotated Example 1 Using the All-LCD Method

Write out the problem leaving plenty of room on both sides and below:

$$\frac{\dfrac{x+2}{x-2} - 4}{\dfrac{1}{x^2-4} + \dfrac{2}{x}}$$

0. Put 4 over 1, and put all polynomials in parentheses:

$$\frac{\dfrac{(x+2)}{(x-2)} - \dfrac{4}{1}}{\dfrac{1}{(x^2-4)} + \dfrac{2}{x}}$$

1. The numerators are already simplified and cannot be factored. All denominators except "x^2-4" cannot be factored. Notice that "x^2-4" is the difference of two squares. Factor it into conjugate pair binomials:

$$\frac{\dfrac{(x+2)}{(x-2)} - \dfrac{4}{1}}{\dfrac{1}{(x-2)(x+2)} + \dfrac{2}{x}}$$

2. Find the LCD of all mini fractions and write it to the side.

    $$LCD = x(x-2)(x+2)$$

3. Multiply all mini fractions by the LCD:

$$\frac{\dfrac{[x(x-2)(x+2)](x+2)}{(x-2)} - \dfrac{4[x(x-2)(x+2)]}{1}}{\dfrac{1[x(x-2)(x+2)]}{(x-2)(x+2)} + \dfrac{2[x(x-2)(x+2)]}{x}}$$

Remove the brackets. Cross out the common factors:

$$\frac{\dfrac{x\cancel{(x-2)}(x+2)(x+2)}{\cancel{(x-2)}} - \dfrac{4x(x-2)(x+2)}{1}}{\dfrac{x\cancel{(x-2)}\cancel{(x+2)}}{\cancel{(x-2)}\cancel{(x+2)}} + \dfrac{2\cancel{x}(x-2)(x+2)}{\cancel{x}}}$$

114

And remove the crossed-out common factors. Notice that all denominators of the mini-fractions will be eliminated (also, remove the denominator "1") and the expression will change from complex to *simple*. It will look like:

$$\frac{x(x-2)(x+2) - 4x(x-2)(x+2)}{x + 2(x-2)(x+2)}$$

4. Simplify both the numerator and denominator, separately, by multiplying, distributing, then combining like-terms
4a. These steps show the multiplication:

$$\frac{x(x^2-4) - 4x(x^2-4)}{x + 2(x^2-4)} = \frac{x^3 - 4x - 4x^3 + 16x}{x + 2x^2 - 8} =$$

4b. This step shows combining like-terms and arranging into descending order:

$$\frac{-3x^3 + 12x}{2x^2 + x - 8}$$

5. Completely factor the numerator and denominator, separately. In the numerator, the GCF is -3x. The denominator is a trinomial, so try to factor it into two binomials.

$$\frac{-3x(x^2-4)}{2x^2 + x - 8} = \frac{-3x(x-2)(x+2)}{2x^2 + x - 8}$$

The denominator cannot be further factored.

Note: the numerator could have automatically been factored to this from step 4a either by using "(x^2-4)" as a GCF, or by combining like-terms, but this will not always be an option.

6. Look for any common factors in the numerator and denominator. In this case, there are non, so the last step is the most simplified form.

Annotated Example 2 Using the Overall Numerator & Denominator Method

Write out the problem and leave plenty of room on the sides of each term:

$$\frac{4 + \dfrac{2}{x}}{\dfrac{x}{4} + \dfrac{1}{8}}$$

0. Put 4 over 1. In this case, there are no polynomials to put parenthesis around in any mini-fraction:

$$\frac{\dfrac{4}{1} + \dfrac{2}{x}}{\dfrac{x}{4} + \dfrac{1}{8}}$$

1. Since there are no polynomials in the numerators, they can't be factored. The denominators in the top fractions can't be factored, however the denominators of the bottom fractions can be factored (into exponential form in anticipation of finding the LCD):

$$\frac{\dfrac{4}{1} + \dfrac{2}{x}}{\dfrac{x}{2^2} + \dfrac{1}{2^3}}$$

2. Find the LCDs:
    a. Find the LCD of the top fractions. It is "x".
    b. Find the LCD of the bottom fractions. It is "8".

3. Convert fractions into like-fractions, then add:
   a. In the overall-numerator, then
   b. In the overall-denominator.
4.
   a. Leave denominator factors un-multiplied.

$$\frac{\frac{(x)4}{(x)1} + \frac{2}{x}}{\frac{(2)x}{(2)2^2} + \frac{1}{2^3}} = \frac{\frac{4x}{x} + \frac{2}{x}}{\frac{2x}{2^3} + \frac{1}{2^3}} = \frac{\frac{(4x+2)}{x}}{\frac{(2x+1)}{2^3}}$$

   b. Simplify all numerators of mini factors. In this case, they are already simplified.
   c. Factor the numerators of both mini-fractions. The numerator of the bottom fraction can't be factored.

$$\frac{\frac{2(x+1)}{x}}{\frac{(2x+1)}{2^3}}$$

5. Divide the top fraction by the bottom fraction by inverting and multiplying:

$$\left(\frac{2(x+1)}{x}\right)\left(\frac{2^3}{(2x+1)}\right)$$

There are no common factors to cancel out, so the simplified form is:

$$\frac{16(x+1)}{x(2x+1)}$$

The Wrong Way to Simplify a Rational Expression

This section highlights a serious mistake that students make all the time. It involves the last step of simplifying a rational expression. Oddly enough, students often perform the more difficult part of the problem correctly before getting to this step, which is why I believe students commit this mistake more out of laziness than ignorance. Regardless of why, it must be prevented, especially because this is often the last step in a problem (and if you have an instructor that doesn't give partial credit, this step could make or break a problem). Here are examples of the wrong and right way to simplify a rational expression. The step(s) I'm highlighting in this section are the same seen in steps 5 & 6 of The All-LCD Method for Simplifying Rational Expressions.

What you need to realize is: you *can't* factor out a *term* in the numerator with a *term* in the denominator when (and because) terms are separated by "+" and "-" signs. You can only cancel *factors* in the numerator with *factors* in the denominator… and *factors* are multiplied, not added or subtracted, together.

The *wrong* thing to do is to instantly cancel out a factor in the numerator with a term or factor in the denominator, without first factoring the numerator (either factoring the GCF out or factoring it into smaller polynomials), and taking into consideration the significance of the plus or minus sign on top between the top terms.

Let's start with this example, which contains a binomial in the numerator and a monomial in the denominator:

$$\frac{12x^2 - 6}{3x}$$

The following is the *wrong way*:

$$\frac{\cancel{12x^2}\ 4x - 6}{\cancel{3x}\ 1} = \frac{4x - 6}{1}$$

in which one attempts to factor 3x out of the $12x^2$ (to 4x) and the 3x (to 1).

Alternatively, the following is *also the wrong way*:

$$\frac{12x^2 - \cancel{6}\,2}{1\,\cancel{3}x} = \frac{12x^2 - 2}{x}$$

in which one attempts to factor 3 out of -6 (to -2) and out of 3x (to 1, leaving x).

Notice how, in each *wrong way* example, the terms incorrectly cancelled out have plus or minus signs in front or behind them. This is the key sign (no pun intended) that should tell you not to cancel out terms.

The Correct Way:

You need to look for a GCF in the numerator, which in this case is 6, and then factor it out. At this point, the common factor of 3 can be cancelled out of the numerator (6) and denominator (3x), as shown:

$$\frac{6(2x^2 - 1)}{3x} = \frac{2\,\cancel{6}(2x^2 - 1)}{1\,\cancel{3}x} = \frac{2(2x^2 - 1)}{x}$$

The expression above on the right can be considered the most simplified form. Compare this answer to the wrong answers from before. As I mentioned in the notes at the end of the All-LCD Method for Simplifying Rational Expressions, the final answer can be shown like this, or by distributing (multiplying) the factor of 2 through the $(2x^2 - 1)$ in the numerator. Since simplification often involves complete factorization and not the reverse (multiplying), I believe this form is the most simplified. If you choose to multiply through (perhaps at the suggestion of your instructors – you should always report the answer the way they prefer it, since they're grading you), it will appear like this:

$$\frac{4x^2 - 2}{x}$$

or, if you break it apart into separate fractions:

$$\frac{4x^2}{x} - \frac{2}{x}$$

which will then simplify to:

$$4x - \frac{2}{x}$$

Again, find out from your instructors how they want you to report your answer.

Let's look at another example, one with a trinomial in the numerator and a binomial in the denominator:

$$\frac{3x^2 - 15x + 18}{2x - 6}$$

The Wrong Way:

There are many wrong ways to approach a such problem. One wrong way might be to attempt to cancel out x from $3x^2$ (to 3x) in the numerator and from 2x (to 2) in the denominator. Another wrong thing to do would be to factor 3 out of -15x (to $-5x$) in the numerator and out of -6 (to -2) in the denominator. If those erroneous cancellations were performed, it would *wrongly* give:

$$\frac{3x - 5x + 18}{2 - 2} = \frac{2x + 18}{0}$$

… which would conclude to be undefined.

The **Correct Way**:

Going back to the original example, factor the GCF (which is 3) out of the numerator. Then, factor the GCF (which is 2) out of the denominator, which would make:

$$\frac{3(x^2 - 5x + 6)}{2(x - 3)}$$

Next, go back to the numerator and see if the trinomial inside parentheses can be factored, which it can be, into the two binomials, seen below. It is revealed that the common factor in the numerator and denominator to be cancelled out is (x − 3), shown below:

$$\frac{3(x - 3)(x - 2)}{2(x - 3)} = \frac{3\cancel{(x - 3)}(x - 2)}{2\cancel{(x - 3)}}$$

After (x − 3) is cancelled out, the final simplified form is: $\frac{3(x-2)}{2}$

Compare this to the wrong answer shown above.

Extraneous Solutions

It is important you know what *extraneous solutions* are, when to look for them, and how to deal with them, because they are tricky and deceptive things. *Extraneous solutions* (also commonly known as *extraneous roots*) appear to be solutions to a problem you just solved, but actually aren't. They tend to come from the following two places:
- A (variable in the) denominator, and
- A (variable inside a) radical.

Based on the location of variables in an equation, these can be thought of as *solution-exceptions*, for the following reasons:
- Any fraction whose denominator is zero is undefined.
- Also, anytime a radicand (of any even root) is negative, the result is not real (still counts as undefined).

You can find extraneous solutions in one of two ways.
- One is by checking all answers after you've solved for the unknowns.
- Another is by finding it (or them) first, *before* solving the equation.
   - I recommend this way, as explained in the next section; it's easier, and this way, if you forget to do the check step at the end of a problem, as many people do, it won't matter. The method for finding extraneous roots in radicals is shown in the Radicals, Roots & Powers section.

Procedure for Solving Equations with Rational Expressions & Extraneous Solutions

A. Find extraneous solutions (solution-exceptions) first [find all possible values of x that would make the denominator (any denominator in the problem) = 0]. When you get to the end of the problem, compare your solutions to the exceptions, and eliminate the *extraneous solutions* from your answers. To do this:

A0. Write out the denominators only (separately, if more than one fraction).

A1. Factor each denominator.

A2. Set all denominator factors = 0 and solve for x (or whatever the variable is).

A3. Put a slash though the = sign, to remind yourself that x *does not equal* the number(s) just determined. Save these off to the side to refer to them at the end of Part B.

B. Solving the Problem:

B0. Write out the whole problem. Write the denominators as the factors you determined through factoring from step A1.

B1. Determine the LCD.

B2. Multiply the LCD times each (numerator only of each) fraction and non-fraction-term (on both sides of the equation). This will eliminate all denominators (and thus all fractions).

B3. Simplify (combine like-terms) and solve using the Procedure for Solving a Simple Algebraic Equation with One Variable).

B4. Compare your answers to those found in step A3 and cross out any extraneous solutions.

Cross Multiplication

Cross Multiplication is the act of multiplying the numerator of one fraction times the denominator of the fraction on the other side of the equal sign, and vice versa.

Cross multiplication is commonly used when doing problems involving proportions, specifically when there is one fraction (only) on each side of the equal sign.

When should you use it? You should use it when trying to solve for a variable in the denominator and when there is only one fraction on each side of the equal sign.

You can only cross multiply if there is only one fraction on each side of the equal sign or you simply can't do it. However...

- If you have more than one fraction on either side of the equal sign, you can either:
  ○ Move one of the fractions to the other side (you can do this if you have two fractions on the same side equal to zero on the other side), or:
  ○ Find the LCD of all fractions & multiply all fractions by the LCD. This will then eliminate all denominators and you will no longer have to do cross-multiplication.

Don't be fooled. For cross multiplication to occur, there must be one fraction on each side of the equal sign, however, the numerators & denominators themselves can be polynomials (if they are, multiply accordingly). Also, you can easily convert a whole number or polynomial into a fraction by putting it over "1".

A mistake students commonly make is trying to cross multiply fractions that are on the same side of the equal sign. Cross multiplication can only be performed *across* equal signs.

See in the example below how:

$\dfrac{2}{3} = \dfrac{5}{x}$ is cross-multiplied to become $(2)(x) = (5)(3)$ which becomes

$2x = 15$ and can be solved by dividing both sides by the coefficient 2:

$\dfrac{2x}{2} = \dfrac{15}{2}$, and thus $x = \dfrac{15}{2}$ or 7.5

Cross-multiplication is not the same as multiplying fractions (on the same side of the equal sign). When fractions are on the same side, multiply the numerators by numerators and the denominators by denominators (see: Multiplying Fractions). Also, Cross-Multiplication is different than "Cross-Cancelling."

Cross-Multiplication vs. Cross Cancelling

It is important to use these methods at the appropriate times and to use the terminology correctly, as they are completely different.

Cross-multiplication is done when you have one fraction set equal another fraction, one of which contains an unknown variable. This is often seen when doing work with proportions and sometimes percent problems. Cross multiplication is and can only be performed *across* an "=" sign by multiplying the numerator of the left fraction by the denominator of the right fraction, and setting that product equal to the product of multiplying the denominator on the left times the numerator on the right. This is explained in more detail in the previous section.

Cross canceling is a simplification technique. It is the process of *simplifying and reducing* fractions by *canceling out* common factors in the numerator of one fraction with the denominator of itself or another fraction it is multiplied by. See in the example below how

$$\left(\frac{4}{15}\right)\left(\frac{5}{6}\right) = \left(\frac{2\cdot 2}{5\cdot 3}\right)\left(\frac{5}{2\cdot 3}\right) \text{ when factored,}$$

which reduces to $\left(\frac{2}{3}\right)\left(\frac{1}{3}\right)$ because

- one of the top 2s cross-cancels with the 2 in the bottom of the other fraction, and
- the 5 in the top cross-cancels with the 5 in the bottom of the other fraction,

which then equals $\frac{2}{9}$ after you multiply the fractions.

RADICALS, ROOTS & POWERS

Note: Before beginning, I must stress (again) that radicals may resemble long division, but they are completely different. Students sometimes mix them up and try to apply long division to radicals, but dealing with radicals is not division. Pay special attention to treat radicals as their own unique function.

Why do you need to understand radicals? One of the main needs and uses is for solving quadratic equations using the Quadratic Formula. They are also commonly applied in problems using the Pythagorean Theorem (which is not covered in this book).

Why is using radicals easy? Because they are nothing more than simple rearrangements of factors. You just have to know what you're looking for. I will tell you what to look for in the coming pages.

When it comes to dealing with radicals, the main objective is to simplify them. To simplify them, you must be able to toggle between different versions of them, and rearrange them. To do that, you must have a good handle on *perfect squares* and factoring.

Perfect Squares & Associated Square Roots

You can take the square root of *any* (positive) number or term, but the result may not be an integer. *Perfect squares* are numbers whose square roots are integers (non-decimal numbers). A set of very common perfect squares are listed after the following explanation.

The list of radicals about to be listed follow the relationship seen below:

$$\sqrt[root]{x^{same \# as \ root}} = \left(\sqrt[root]{x}\right)^{same \# as \ root} = x.$$

In words: When the root of a radical is the same as the power of the base of the radicand, the radical simplifies to that base (which, here, is "x"). Also, if a radical is raised to the same number as the root, it also simplifies to equal the base (again, here, is "x"). The important thing to realize is that *both* simplify to the same base, "x," in this case. This equality also exemplifies another property, which is that the exponent can be moved from the radicand to outside the radical, and vice versa. This is a necessary manipulation technique.

Observe this example using 7 as the same root and power:

$$\sqrt[7]{x^7} = \left(\sqrt[7]{x}\right)^7 = x$$

In words: The seventh root of base x to the power of seven equals x; and: The seventh root of x in parentheses, raised to the power of seven (outside the parentheses) also equals x.

Notice in both versions, the root is the same as the exponent, and both versions equal "x". You will notice this trend in the list of perfect squares and square roots, in the next pages.

This also follows in the next example where the root and power are both 2. You may notice there is no "2" written as the root, but don't let that deceive you. For square roots, the root 2 is usually not written, but it's also not wrong if you write it in.

$$\sqrt{x^2} = (\sqrt{x})^2 = x$$

In words: The square root of a radicand whose base is squared equals that base.

Equivalently: The square of the square root of some radicand equals that radicand.

Notice: whether the square is inside the radical or outside the radical, the result is the same.

Before working with (simplifying) radicals, it is important to know some of the common perfect squares. In the next list, the left column shows the squares, and to the right shows the associated square roots. Having these memorized will help you simplify radicals more quickly and easily. These are listed as an easy reference, but also so you can see the patterns as discussed above.

List of Perfect Squares & Associated Square Roots

$0^2 = 0;$ $\sqrt{0} = \sqrt{0^2} = (\sqrt{0})^2 = 0$

$1^2 = 1;$ $\sqrt{1} = \sqrt{1^2} = (\sqrt{1})^2 = +/-1$

$2^2 = 4;$ $\sqrt{4} = \sqrt{2^2} = (\sqrt{2})^2 = +/-2$

$3^2 = 9;$ $\sqrt{9} = \sqrt{3^2} = (\sqrt{3})^2 = +/-3$

$4^2 = 16;$ $\sqrt{16} = \sqrt{4^2} = (\sqrt{4})^2 = +/-4$

$5^2 = 25;$ $\sqrt{25} = \sqrt{5^2} = (\sqrt{5})^2 = +/-5$

$6^2 = 36;$ $\sqrt{36} = \sqrt{6^2} = (\sqrt{6})^2 = +/-6$

$7^2 = 49;$ $\sqrt{49} = \sqrt{7^2} = (\sqrt{7})^2 = +/-7$

$8^2 = 64;$ $\sqrt{64} = \sqrt{8^2} = (\sqrt{8})^2 = +/-8$

$9^2 = 81;$ $\sqrt{81} = \sqrt{9^2} = (\sqrt{9})^2 = +/-9$

$10^2 = 100;$ $\sqrt{100} = \sqrt{10^2} = (\sqrt{10})^2 = +/-10$

$11^2 = 121;$ $\sqrt{121} = \sqrt{11^2} = (\sqrt{11})^2 = +/-11.$
You get the idea…

$12^2 = 144$ $13^2 = 169$

$14^2 = 196$ $15^2 = 225$

$16^2 = 256$ $17^2 = 289$

$18^2 = 324$ $19^2 = 361$

$20^2 = 400$

Common Perfect Cubes & Associated Cube Roots

The following is a list of common perfect cubes and their associated cube roots. They are here for you to reference, but you should also memorize them and notice the patterns of moving the exponent in and out of the radical as discussed a few pages ago.

$0^3 = 0;$ $\sqrt[3]{0} = \sqrt[3]{0^3} = (\sqrt[3]{0})^3 = 0$

$1^3 = 1;$ $\sqrt[3]{1} = \sqrt[3]{1^3} = (\sqrt[3]{1})^3 = 1$

$-1^3 = -1;$ $\sqrt[3]{-1} = \sqrt[3]{-1^3} = (\sqrt[3]{-1})^3 = -1$

$2^3 = 8;$ $\sqrt[3]{8} = \sqrt[3]{2^3} = (\sqrt[3]{2})^3 = 2$

$3^3 = 27;$ $\sqrt[3]{27} = \sqrt[3]{3^3} = (\sqrt[3]{3})^3 = 3$

$4^3 = 64;$ $\sqrt[3]{64} = \sqrt[3]{4^3} = (\sqrt[3]{4})^3 = 4$

$5^3 = 125;$ $\sqrt[3]{125} = \sqrt[3]{5^3} = (\sqrt[3]{5})^3 = 5$

$6^3 = 216$ $7^3 = 343$

$8^3 = 512$ $9^3 = 729$

$10^3 = 1000$

Other Powers & Relationships of 2, 3, 4 & 5

This is an extra section to show other common powers and exponential relationships for base 2 through base 5, at powers of 4 and 5. Notice here how these numbers can be factored and rearranged using the rules of multiplying bases with exponents and taking powers of powers.

$2^4 = (2^2)(2^2) = (2^2)^2 = (4)(4) = 16$

$3^4 = (3^2)(3^2) = (3^2)^2 = (9)(9) = 81$

$4^4 = (4^2)(4^2) = (4^2)^2 = (16)(16) = 256$

$5^4 = (5^2)(5^2) = (5^2)^2 = (25)(25) = 625$

$2^5 = (2^2)(2^3) = (2^2)^3 = (2^3)^2 = 32$

$3^5 = (3^2)(3^3) = (3^2)^3 = (3^3)^2 = 243$

$4^5 = (4^2)(4^3) = (4^2)^3 = (4^3)^2 = 1024$

$5^5 = (5^2)(5^3) = (5^2)^3 = (5^3)^2 = 3125$

These lists are here for your reference. Keep these power and root relationships in mind for the next section, as they play a helpful role in manipulating and simplifying radicals.

Manipulating & Simplifying Radicals

The reason it's important to be able to recall the perfect squares, perfect cubes, and other perfect powers is because they are essential in simplifying radicals, which is why I listed many of them in the previous section. One major reason for simplifying radicals is to find which radicals are like-terms, so they may be combined as you would combine like-terms with variables (and there are other reasons, too).

It is important to know that simplifying radicals is different than simplifying other terms or expressions that don't have radicals, so you can't expect to use the same strategy. The main difference is that when simplifying non-radical terms or expressions, you usually resort to factoring into prime factors and/or finding a GCF to factor out. Simplifying radicals actually means factoring and *reorganizing* factors of the radicand, but the radicand is not necessarily prime factored.

To simplify radicals, you must factor the radicand into two types of factors:
- perfect-power-factors and
- non-perfect-power factors.

And there is a very logical reason for this. The radical of the perfect-power is to be taken, and then (its root) will be moved outside the radical (and treated like a coefficient that is multiplied by the remaining radical). The non-perfect-power factor will simply remain under the radical because that is its most simplified form.

Observe this method in the next section Common Radical Fingerprints. Look at the example for the square root of 12. Notice that its factor "4" is a perfect square factor, and "3" is not, so you separate them into factors (4)(3). Now, since 4 is a perfect square, take the square root of it. Since the square root of 4 is 2, the 2 gets moved to the outside of the radical as a coefficient, and the square root of 3 remains in the radical, leaving you with (as you would say) "two (times) the square root of three."

The reason radicals are simplified this way is so they can be manipulated into like-terms that can be combined (as in "combine like-terms").

For radicals, "like-terms" are terms in which both the root and radicand are exactly the same. When these criteria are met, like-radicals are combined via their coefficients the same way as like-terms with variables.

As in the last section in which I show a list of common roots and powers, there are others that are still common, but "not perfect"...
"not perfect" in the sense that the radical cannot be reduced to an integer. These are so common, that I call them "fingerprints," because after encountering them enough, you may memorize them, saving you the step of having to manually factor and simplify them every time.

The following list accomplishes three purposes.
1. It simply shows common "non-perfect" radicals, and
2. it shows the intermediate steps where the radicands are factored into "perfect-powers" (in this case, they're perfect squares), and "non-perfect powers" (which in this case are non-perfect squares).
3. It also demonstrates the "Product Rule of Radicals."

I also decided to include a few common radicals which are already in their most reduced form, just to put them into perspective.

List of Common Radical Fingerprints

$$\sqrt{8} = \sqrt{(4)(2)} = \sqrt{4}\sqrt{2} = +/- 2\sqrt{3}$$

$$\sqrt{10} = +/-\sqrt{10}$$

$$\sqrt{12} = \sqrt{(4)(3)} = \sqrt{4}\sqrt{3} = +/-2\sqrt{3}$$

$$\sqrt{18} = \sqrt{(9)(2)} = \sqrt{9}\sqrt{2} = +/-3\sqrt{2}$$

$$\sqrt{20} = \sqrt{(4)(5)} = \sqrt{4}\sqrt{5} = +/-2\sqrt{5}$$

$$\sqrt{24} = \sqrt{(4)(6)} = \sqrt{4}\sqrt{6} = +/-2\sqrt{6}$$

$$\sqrt{27} = \sqrt{(9)(3)} = \sqrt{9}\sqrt{3} = +/-3\sqrt{3}$$

$$\sqrt{28} = \sqrt{(4)(7)} = \sqrt{4}\sqrt{7} = +/-2\sqrt{7}$$

$$\sqrt{30} = +/-\sqrt{30}$$

$$\sqrt{32} = \sqrt{(16)(2)} = \sqrt{16}\sqrt{2} = +/-4\sqrt{2}$$

$$\sqrt{40} = \sqrt{(4)(10)} = \sqrt{4}\sqrt{10} = +/-2\sqrt{10}$$

$$\sqrt{44} = \sqrt{(4)(11)} = \sqrt{4}\sqrt{11} = +/-2\sqrt{11}$$

$$\sqrt{45} = \sqrt{(9)(5)} = \sqrt{9}\sqrt{5} = +/-3\sqrt{5}$$

$$\sqrt{48} = \sqrt{(16)(3)} = \sqrt{16}\sqrt{3} = +/-4\sqrt{3}$$

$$\sqrt{50} = \sqrt{(25)(2)} = \sqrt{25}\sqrt{2} = +/-5\sqrt{2}$$

$$\sqrt{51} = \sqrt{(3)(17)} = +/-\sqrt{51}$$

$$\sqrt{52} = \sqrt{(4)(13)} = \sqrt{4}\sqrt{13} = +/-2\sqrt{13}$$

$$\sqrt{54} = \sqrt{(9)(6)} = \sqrt{9}\sqrt{6} = +/-3\sqrt{6}$$

$$\sqrt{56} = \sqrt{(4)(14)} = \sqrt{4}\sqrt{14} = +/-2\sqrt{14}$$

$$\sqrt{60} = \sqrt{(4)(15)} = \sqrt{4}\sqrt{15} = +/-2\sqrt{15}$$

$$\sqrt{63} = \sqrt{(9)(7)} = \sqrt{9}\sqrt{7} = +/-3\sqrt{7}$$

$$\sqrt{68} = \sqrt{(4)(17)} = \sqrt{4}\sqrt{17} = +/-2\sqrt{17}$$

$$\sqrt{75} = \sqrt{(25)(3)} = \sqrt{25}\sqrt{3} = +/-5\sqrt{3}$$

$$\sqrt{76} = \sqrt{(4)(19)} = \sqrt{4}\sqrt{19} = +/-2\sqrt{19}$$

$$\sqrt{80} = \sqrt{(16)(5)} = \sqrt{16}\sqrt{5} = +/-4\sqrt{5}$$

$$\sqrt{84} = \sqrt{(4)(21)} = \sqrt{4}\sqrt{21} = +/-2\sqrt{21}$$

$$\sqrt{88} = \sqrt{(4)(22)} = \sqrt{4}\sqrt{22} = +/-2\sqrt{22}$$

$$\sqrt{90} = \sqrt{(9)(10)} = \sqrt{9}\sqrt{10} = +/-3\sqrt{10}$$

$$\sqrt{98} = \sqrt{(49)(2)} = \sqrt{49}\sqrt{2} = +/-7\sqrt{2}$$

$$\sqrt{99} = \sqrt{(9)(11)} = \sqrt{9}\sqrt{11} = +/-3\sqrt{11}$$

$$\sqrt{104} = \sqrt{(4)(26)} = \sqrt{4}\sqrt{26} = +/-2\sqrt{26}$$

$$\sqrt{117} = \sqrt{(9)(13)} = \sqrt{9}\sqrt{13} = +/-3\sqrt{13}$$

$$\sqrt{120} = \sqrt{(4)(30)} = \sqrt{4}\sqrt{30} = +/-2\sqrt{30}$$

$$\sqrt{124} = \sqrt{(4)(31)} = \sqrt{4}\sqrt{31} = +/-2\sqrt{31}$$

$$\sqrt{125} = \sqrt{(25)(5)} = \sqrt{25}\sqrt{5} = +/-5\sqrt{5}$$

It is important to note that when doing square roots (or any even roots) on a calculator, most calculators will only report the positive root, so it is up to you to also write the negative.

Extraneous Roots in Radical Equations

The topic of extraneous roots (a.k.a. extraneous solutions) has been explained previously, as well as how to identify them when they are in a denominator. But they can also be in an equation with radicals. When they are, you must check your answer by substituting back into the original equation and simplifying. This is mentioned in an upcoming section: Checking Your Answers.

FMMs (FREQUENTLY MADE MISTAKES)

This section of the book is one like no other. I bet you will not find one like this in a traditional textbook (at least I haven't yet). This section is strictly dedicated to highlighting the most common and "frequently made mistakes" by students. By honing in on these common mistakes, I hope you will be able to quickly recognize and avoid them.

This section is also different than the other sections in *this* book in the way it is set up. If it's a topic that hasn't been covered in the book yet, I will give it its own new section. If it's a mistake I've already explained in a previous section, I will list it with a brief introduction and provide the page number to direct you back to that respective section of the book.

The Two Meanings of "Cancelling Out"

"Cancelling out" can mean two different things:
1. Cancelling out to *zero*, or
2. Cancelling out to *one*.

Students and instructors often just say "cancelling out," which is a bit ambiguous and can cause confusing. You must be able to properly differentiate which context is being used and when each is happening.

Terms are *cancelled to zero* when *opposite* terms are added (meaning adding and subtracting the same term. This is often seen when you are adding or subtracting the same term to each side of the equation in order to *move* a term to the opposite side. It is also seen during "combine like-terms," when terms happen to be *opposites* of each other.

Terms *cancel out to one* when:
- A number or term is divided by itself, or
- A fraction is multiplied by its reciprocal.

This is often seen during:
- Reducing fractions;
- Multiplying fractions;
- The final step of solving a simple algebraic equation of one variable, where you divide both sides by the coefficient in front of the variable you're solving for; and
- Factoring a GCF out of a series of terms.

Students usually don't have problems remembering that adding opposites cancels them to zero. But sometimes the mistake is when students think canceling *always* results to zero. You must not forget that when a term is divided by itself, it equals "1," as shown in Property Crises of Zeros, Ones and Negatives. This typically involves fractions (either during reducing individual fractions or multiplication or division of fractions).

But the most common time it is forgotten is when you find a GCF in a series of terms and then factor that GCF out (by dividing each term by it), leaving the GCF out front, multiplied by the parentheses containing the remaining factors of each term. For example, in the following expression:

$$18x^3 - 6x^2 + 3x$$

the GCF is 3x. To simplify this, you would factor 3x out of each term. The intermediate step (which you wouldn't always show) shows each term divided by 3x:

$$\frac{18^3}{3x} - \frac{6x^2}{3x} + \frac{3x}{3x}$$

Students often *incorrectly* answer this as:

$$= 3x(6x^2 - 2x).$$

When asked about it, they will respond that
"three x over three x cancels out," which is true, but it cancels to "1," not zero, so the 1 must be shown, as in the correct answer shown here:

$$= 3x(6x^2 - 2x + 1)$$

While on the subject of "cancelling out," this plays a role in multiplication and division of fractions by means of "cross cancelling." Another common mistake or area of confusion is when students mix up cross cancelling with cross multiplication. This is explained in: Cross Multiplying vs. Cross Cancelling.

Miscellaneous Mistakes

- Sometimes students are required to distribute an exponent through a term of multiple (variable) bases with exponents. This is taking the power of each base to the power being distributed. There is often a coefficient attached to the variables, and when there is, students often forget to apply the power (from outside the parentheses) to the coefficient. The reason might be because students are used to taking the power of the power of each variable base, and they just forget about the coefficient because coefficients rarely have a written exponent... because it's usually to the unwritten power of 1. When distributing an exponent through a group of bases with a coefficient, don't forget to apply the exponent to the coefficient.

- Students often make a mistake when a negative sign is in front of a fraction by not properly distributing the negative sign through, changing the sign of each term in the series.

- When multiplying factors of a common base with exponents, sometimes students mistakenly multiply the exponents. When factors of a common base are multiplied, their exponents are added. See page 40.

- When given an equation with a trinomial, or a quadratic equation, sometimes students will successfully factor it, but then forget to do the last step, which is to solve. Don't forget to solve by setting each factor to zero, then solving for the variable. A factor may be a set of parentheses or an already factored out variable. This step is demonstrated through various examples in the Trinomials & Quadratic section, starting on page 86.

- Students commonly make the mistake of using "zero" and "no slope" or "undefined" interchangeably, but they have completely different meanings. See:
  o What Does Undefined Mean? page 49
  o When $x_1 = x_2$ and When $y_1 = y_2$, page 54;

- For common mistakes students make during the Substitution Method for solving a system of two linear equations, see page 62.

- For common mistakes students make during the Addition/Elimination Method for solving a system of two linear equations, see page 64.

- Equations and expressions are intended to be simplified completely. Often times, students do most of the problem correctly, but make one of two vital mistakes that could make or break an answer (especially when instructors don't give partial credit). Sometimes students get near the end, but simply forget to simplify the answer. Or, sometimes students attempt to simplify, but do it wrong. Learn to avoid: The Wrong Way to Simplify a Rational Expression on page 119.

- When students use the Quadratic Formula, they often forget to simplify the last step. This is explained in: The Part Everyone Forgets (The Last Step of the Quadratic Equation), page 83.

- When applying the "special case" shortcut method to multiplying out a binomial squared, students often make the common mistake of using the shortcut method for multiplying conjugate pair binomials. This mistake results in the missing "bx" term. See page 101.

- When a negative sign is in front of a rational expression (a fraction with a polynomial in the numerator) students very often forget to distribute that negative sign through all terms in the numerator. This then incorrectly associates the negative to only the first term in the numerator, leaving the terms to follow with opposite signs than what they should be.

- Radicals Are Not Long Division. There's not really much to say about this other than that the symbols and set up of radicals and long division are similar looking, but they are completely different operations. Anytime I've ever encountered a student attempting to apply long division to a radical may have been their desperate attempt to do something when they had no idea how to approach radicals (most likely due to lack of preparation). Long division is a process to find out how many times the divisor goes into the dividend, and the answer is the quotient. But radicals are used to answer: What number, which when taken to the power shown as the root, equals the radicand? The radicand won't always be a perfect power number, and in that case, assuming you don't use a calculator, you will break it down and simplify it using the rule of multiplication of roots, as briefly shown in Common Radical Fingerprints on page 136 - 137.

Scientific Notation on Your Calculator

Scientific notation is a standardized way of reporting numbers that are either very big or very small, with many zeros and/or decimal places. It is a way to express numbers into a manageable format, and is often used in science and statistics. Scientific notation is the alternate way of writing a number from its expanded form.

Although I do not cover scientific notation in this book, I want to address the mistake students frequently make when putting scientific notation into a calculator. The mistake is some variation of not knowing how to properly put it into the calculator.

Since there are generally two types of calculators (scientific and graphing) with the scientific notation function, each type and brand varies in what buttons they have to accomplish this function, so it's a good idea to be prepared for each possibility. There is also a completely *wrong* way to input scientific notation, which results in the number being off by an order of magnitude (a factor of 10, or in other words, off by one zero).

Typically, on all calculators, you start off the same, by typing in the base number. Next, you must hit the exponent button, but not the same exponent button you would use for normal exponents. The button you want may look like any of the following:

[EE], [exp], [EXP], [x 10] (meaning "times ten to the ...")

$[10^x]$ (meaning " base ten to the power of")

[anti LOG] (often a *2nd function* to [LOG])

[e] (not to be confused with $[e^x]$, which stands for "the number e to the x," also known as "inverse LN," which is "inverse natural log").

If one of the functions shown above is a *2nd function*, meaning the symbol is shown in another color, above a primary button, you must hit or hold a button such as: "2nd," "Shift," or "Alt," often located at the top left corner of the calculator, then hit the button as it is shown (from the choices listed above). You might have to look around for it; it doesn't always jump out at you at first.

To reiterate, you would first type your base number, then scientific notation button (shown above), then the exponent (of the 10).

Here is the place students often make a mistake… by manually typing out:

[the base #] [x] [10] [EXP] [the exponent]… In words, that would say, "base number times ten, times ten to the power of some number." In other words, this causes a redundant multiple of 10, which will result in your number being off by a factor of ten. To prevent this, you must use either one or the other:

- either [x 10 ^ the exponent], or

- the [EXP, then the exponent]

Consider the example of converting 9,400,000 to scientific notation, which would be

9.4×10^6. You would type $[EXP][6] = [\times][10][^][6]$

- Either: [9.4][x][10][^][6]
In words: Nine point four times ten to the sixth, using the exponent feature, not the scientific notation feature.

- Or: [9.4][EXP][6]
This is the preferred way to input numbers in scientific notation. In words, this reads the same as above ("Nine point four times ten to the sixth"), but the buttons are clearly different. In this version, the scientific notation button is used, not manually typing the ten, the carrot (^), and the exponent six. I recommend getting use to the [EXP] button.

↑ scientific notation button

143

What Does "Error" on a Calculator Mean?

Often times, students will put an operation into the calculator and get the response: "Error." Some misinterpret what that means. Sometimes students interpret that as "the student made an error," but this many not be the case. When the output on the calculator is "Error," it could mean one of the following things:

1. "Error" is the correct and expected response. What we might call "Undefined" or "No Solution" to an arithmetic operation, the calculator will report as "Error." For such examples involving division and radicals, see: Property Crises of Zeros, Ones and Negatives).

2. Sometimes, however, "Error" means you made a mistake in-putting your intended operation. In that case, you should check what you typed and look for an error in that respect. For instance, you may have accidentally typed two decimals in a number. If you check, and don't find an input mistake, then there is a good chance "error" is the correct response for a reason.

144

Checking Your Answers

The last step of any problem-solving procedure is to check your answer. Specifically, that means taking the value of the unknown you determined, substituting it back into the original equation, simplifying, and reviewing the outcome. If, after simplifying, the left side equals the right side, this affirms that your answer is correct. However, if it doesn't, there are three possible reasons why.

1. You made an error doing the problem.
2. You've identified an extraneous solution.
3. You made an error in your math in the check step.

One clear reality is that students either forget, or just hate doing the check step. If I had to guess, it's because it costs extra time and workspace to do, and students just want to be done with a problem, especially problems that are long to begin with. But to attain complete answers, you're expected to check your answers, and there *are* times when checking will help you discover an error or anomaly. This could save you valuable points on a test.

First, it will simply draw your attention to an error you made during problem solving. If you can't find your mistake by reviewing your work, consider starting it over without looking at the last way you did it. Consequently, you can also make an error during the check steps which may lead you to think you made an error in the original steps, but didn't. Either way, your answers should check out.

There's also another major reason an answer might not check out, and that is due to an extraneous solution. Extraneous solutions often occur when a variable is in the denominator in an equation (or inside a radical).

CLOSING

My mission is to help average people break through math barriers, whatever the source of the barriers may be. You learn based on how you are taught. I learn how to teach based on the trials, errors and patterns of what and how students learn. It is a constant learning curve trying to perfect how to predict how students process the lessons; the learning is a two-way street. I am extremely interested in your feedback. Please tell me what worked, what you liked, what you didn't like, what was confusing, and what you'd like to see (more of). If this book helped you, I ask that you please support my mission by telling your friends and family about it. I wish you the best of luck in all you do.

Please send your feedback to my personal email address: bullockgr@gmail.com.

Feel free to follow @GregBullock and @AlgebraInWords on Twitter.

CPSIA information can be obtained at www.ICGtesting.com
Printed in the USA
LVOW04s0801241114

415296LV00001B/53/P